Rock Mechanics with Emphasis on Stress

Rock Mechanics with Emphasis on Stress

Editor

Fritz Rummel

CRC Press
Taylor & Francis Group
Boca Raton London New York

CRC Press is an imprint of the
Taylor & Francis Group, an **informa** business

A BALKEMA BOOK

CRC Press
Taylor & Francis Group
6000 Broken Sound Parkway NW, Suite 300
Boca Raton, FL 33487-2742

First issued in hardback 2019

ISBN-13: 978-0-415-37465-1 (hbk)

Visit the Taylor & Francis Web site at
http://www.taylorandfrancis.com

and the CRC Press Web site at
http://www.crcpress.com

Preface

The formulation of the plate tectonics concept during 1960 to 1970 is considered as a revolution in earth sciences and initiated a renaissance to various subjects related to geological structures and processes. It is important, both, for the understanding of long-term earth deformation behaviour and for short-term deformation phenomena such as earthquakes or the stability of geo-engineering structures. At that time, I was a young researcher at the University of Minnesota, USA, and participated—to a small extent—in the emergence of Rock Mechanics from more or less empiric technical approaches into a scientific discipline with both application to geodynamics and practical geo-engineering. It was the time when J.C. Jaeger and N.G.W. Cook published the bible on Fundamentals of Rock Mechanics, when Jaeger often visited Minneapolis and silently observed my rock fracture experiments with the first servo-controlled testing machine, when Charles Fairhurst encouraged H. von Schoenfeldt to carry out in-situ hydraulic fracturing tests for first reliable crustal stress measurements. Since then, Rock Mechanics has further developed into one of the areas of Geotechnology to attack the geo-engineering challenges of the next decades. Rock Mechanics will contribute to the understanding of earthquake mechanics, to the development of geothermal as a future energy option, to guarantee fresh water supply, to a safe storage of hazardous wastes, and to large scale underground transport.

Rock Mechanics, to me, is the study of the response of rock to the force field at a given environment. The study assesses the relationship between the strength of crustal rock masses on different scales, and the acting tectonic or artificially imposed stresses, whether it is the aim to predict, prevent, or promote the collapse of geological or engineering structures. To our present knowledge, we may accept that the state of stress in the Earth's upper crust is in equilibrium with the frictional resistance of the rock. Thus, any change of stresses caused by tectonic processes or induced by engineering actions may lead to fracture deformation or catastrophic failure. Therefore, the knowledge of stress is of utmost interest to both, earth scientists and to geo-engineers.

On a global scale, crustal stresses are the result of plate tectonics. Enormous efforts were recently undertaken in the World Stress Map project of the International Lithosphere Programme to compile stress data worldwide and to correlate continental stress patterns with earthquake activities. We now recognize that intraplate earthquakes or earthquakes in stable continental regions, like on the Indian subcontinent, may be common events. On a regional scale, the knowledge of stresses is of vital interest for deep mining like in Chuquicamata (Chile), geothermal heat exploitation for electricity generation from deep hot rock formations (HDR concept), safe underground hazardous waste disposal scenarios in impermeable rock, large-scale tunnelling for high-speed traffic, and for urban underground mass transport systems.

Many more problem areas could be identified. It was the purpose of the MeSy (India) first workshop on Rock Mechanics with Emphasis on In-situ Stress to present areas where stress is an important parameter and which presently seems to be of particular interest for the next decades, such as to understand more about large earthquake focal mechanics, to further develop the "Hot Dry Rock (HDR) concept for geothermal energy exploitation, to develop strategies for the stimulation of deep-seated fresh water reservoirs, and to introduce systematic stress measurements in large-scale tunnelling for mass transportation projects and safe underground storage of gas and hazardous waste. The workshop was held at New Delhi in the inciting ambiente of the colonial style Hotel Imperial on 28th and 29th September 2001. The presentations were made by internationally known experts mostly from Germany and India. The present book includes most of the presentations but is also supplemented by articles on the Gujarat earthquake and on Koyna dam induced seismicity. It was exciting to listen to the presentations and to the vivid discussions during the workshop, and—I hope—the reading of this book will stimulate thinking on the fascinating subject of geo-engineering and geoscience. I am very grateful to all the Workshop speakers and to all the authors who made important contributions to this book. I am also grateful to Mr. A.K. Rai, Director of MeSy (India) who organized the Workshop, and to the Central Mining Research Institute, Dhanbad for the support of the Workshop. Mr. Mohan Primlani from the Oxford & IBH Publishing company encouraged me for the publication of the Workshop papers and sometimes pushed me when I got tired to chase after the authors contributions.

December 2003   **Fritz Rummel**

Contents

The Contributors

Hans-Joachim Alheid graduated as geophysicist at the Ruhr-University Bochum (Germany) in 1975 and received his Doctoral Degree in 1982 on Laboratory Earthquake Physics. Since 1981, he is a scientist at the Federal Institute of Geosciences and Natural Resources (BGR), Hannover (Germany). The core area of his research is the detection and characterization of excavation damaged zones around possible nuclear waste disposal sites as Head of the BRR Soil Mechanics/Engineering Seismology Section. E-mail: h-j.alheid@bgr.de

Georg Boor is Head of the Gas Storage Section of Pipeline Engineering GmbH Essen (Germany), a sister company of Ruhr Gas AG Essen. He is involved in the design of numerous large-scale underground gas storage projects all over the world. E-mail: georg.boor@ple.de

Rajender Kumar Chadha has an MSc in Geology and a PhD in Applied Geophysics and is scientist at the National Geophysical Research Institute (NGRI), Hyderabad (India). His main interests are pore pressure studies related to the Koyna reservoir induced seismicity, and broad-band seismological studies in the Peninsular shield of India. On these topics he has published 28 papers and co-edited 2 volumes on *Induced Earthquakes and the Nature of Seismic Sources* and *Prediction of Earthquakes*. E-mail: chadha@ngri.res.in

Michael Alexander Ellis received his PhD in 1984 from Washington State University (USA) for studies of the kinematics of deformation in greenshist grade rocks. Then he concentrated on various aspects of active crustal tectonics, and in 1990 he joined the Center of Earthquake Research and Information at the University of Memphis (USA) where he is presently Director of the Active Tectonics Laboratory. His research is divided between landscape evolution in the face of active tectonics and climatic processes, and in the origin and significance of large non-plate boundary earthquakes.

Address: Center for Earthquake Research and Information, The University of Memphis, Memphis, TN 38152, USA.

E-mail: ellis@ceri.memphis.edu

Thirumandas Narayan Gowd got his Geophysics degree at the Osmania University Hyderabad (India), did his doctoral work with F. Rummel at the Ruhr-University Bochum (Germany), and received his PhD at the Osmania University Hyderabad in 1973. From 1975 to 2002 he was Head of the Rock Mechanics Section of the National Geophysical Research Institute (NGRI), Hyderabad. He became nationally and internationally known through his participation in the World Stress Map Project of the International Lithosphere Programme, by numerous hydraulic fracturing stress measurements in India, and by his summary on the state of stress on the Indian subcontinent. He retired in 2001 and lives in Hyderabad.

Address: Scientist 'G' (Rtd), National Geophysical Research Institute, Hyderabad-500 007, India.

E-mail: tngowd@yahoo.co.in

Gerd Klee did his diploma in Geophysics with F. Rummel at the Ruhr-University Bochum (Germany) in 1990. He joined MeSy GmbH in 1991 and became its co-director in 1994. With MeSy he carried out hydraulic fracturing stress measurements for large-scale engineering projects in Germany, Switzerland, Italy, Belgium, France, England, Sweden, Canada, Hong Kong, India, Kenya, Laos, and Thailand. A special highlight was his participation in stress measurements and induced seismicity in the 9 km deep borehole of the German Deep Continental Drilling project. He has published 13 scientific papers.

E-mail: mesy-bochum2004@t-online.de

Heinz Konietzky received his Doctoral Degree in Geotechnics at the Technical University Freiberg (Germany) in 1989. There, he also did his Habilitation and earned the title Private Dozent 2001. From 1990 to 1993 he was scientist with MeSy GmbH, Germany and sampled experience in hydraulic fracturing stress testing and its interpretation during projects in Hong Kong and Kenya. Since 1993 he is Director of Itasca GmbH at Gelsenkirchen (Germany). He has published more than 50 papers, in geomechanics and on numerical modelling.

E-mail: hkonietzky@itasca.de

Michael Krieter got his Diploma in Geophysics and his Doctoral Degree at the Technical University Berlin (Germany) in 1992. Since 1993, he is a member of the Gas Storage Section of Pipeline Engineering GmbH Essen (Germany) and is responsible for geophysical prospection interpretation and thermodynamic and rock mechanic modelling for the design and operation of gas storage projects in Germany and abroad.

E-mail: michael.krieter@ple.de

Hans-Joachim Kümpel got his Diploma and Doctoral Degree at the University of Kiel (Germany) and became Professor of Geophysics at the University of Bonn (Germany) in 2001. Since July 2001, he is the Director of the Leibniz Institute for Applied Geosciences at Hannover (Germany). His present major interests are on poroelasticity and induced seismicity which resulted during his collaboration with the National Geophysical Research Institute (NGRI) Hyderabad (India) on reservoir induced seismicity in the Koyna-Warna region. (see: www.gga-hannover.de)

E-mail: h.kuempel@bgr.de

Sabrina Leonardi is a geologist from the University of Bologna (Italy). She did her Doctoral Degree under HJ Kuempel at the University of Bonn (Germany) in 1996. She worked as Editor of the Bulletin of the German Geophysical Society. Her major research interests are magma dynamics of Mt. Etna and reservoir induced hydrogeology.

Ian McFeat-Smith graduated as Engineering Geologist and did his PhD on the performance of tunnelling machines. Since 1978, he worked as a qualified geotechnical, civil and mining engineer for major geo-engineering consultants in Asia on the planning and design of over 90 tunnel contracts in Hong Kong, Singapore, Athens, in Jakarta, and in Taipei. In 1997, he launched his own company IMS Tunnel Consultancy in Hong Kong.

Address : Managing Director, IMS Tunnel Consultancy Ltd, Hong Kong

E-mail: ims.tunel@netvigator.com

Fritz Rummel studied Geophysics at the University of Munich (Germany) and received his Doctoral Degree in 1967. From 1968 to 1970 he worked with Professor Charles Fairhurst at the University of Minnesota (USA) where he installed the first servo-controlled testing machine in Rock Mechanics and did the first in-situ hydraulic fracturing stress measurements together with von Schoenfeldt. As Professor of Geophysics at the Ruhr-University of Bochum (Germany) he further developed hydraulic fracturing, concentrated on rock deformation under ultra-high pressure and temperature laboratory conditions, and co-established the European Hot-Dry-Rock project at Soultz-sous-Forêts for geothermal energy exploitation. In 1984, he founded MeSy GmbH Bochum, presently he initiates the project Prometheus for geothermal heat supply to the Ruhr-University Bochum.

Address: MeSy GEO-Meßsysteme GmbH, Meesmannstr. 49, D-44807 Bochum, Germany

E-mail: fritz.rummel@ruhr-uni-bochum.de,   mesy-bochum2004@t-online.de

Amalendu Sinha is presently the Head of the Geomechanics and Mine Technology Section of the Central Mining Research Institute (CMRI), Dhanbad (India). Within India he is a well-known earth scientists in the field of stress measurements, mining technology, and mine safety.

Address: Central Mining Research Institute, Dhanbad, India

E-mail: amalendusinha@hotmail.com

Kirti Srivastava has obtained his MSc and PhD in Statistics. He is a scientist at the National Geophysical Research Institute (NGRI), Hyderabad (India). His main interests are on groundwater modelling. He has published 16 papers.

E-mail: director@ngri.wipro.net.in

Rama Krishna Tiwari graduated from Banares Hindu University and is a scientist at the National Geophysical Research Institute (NGRI), Hyderabad (India) since 1976. In 1985, he earned his PhD. He often visited research institutes in Germany and England, and has specialized in time series in geosciences, catastrophe theory, and non-linear processes. Besides 50 publications he has also received research awards from the Indian Geophysical Society.

E-mail: director@ngri.wipro.net.in

Jan Richard Weber graduated as a Petroleum Engineer from the Technical University Clausthal (Germany) in 1991, where he also received his Doctoral Degree in 1994 on the permeability behaviour of rocks under stress and temperature conditions. Since 1996, he is working as a scientist at the Federal Institute of Geosciences and Resources (BGR), Hannover (Germany) and works on the design, performance, and analysis of hydraulic tests in rock salt. Presently, he is delegated to the German Federal Ministry of Economics, Nuclear Energy Section.

Address: Federal Institute for Geosciences and Natural Resources, Hanover, Germany,

E-mail: jan.weber@bgr.de

The Imperial, built in 1931 on the prestigious Queensway – today Janpath (Peoples' Boulevard)—with its Victorian/Colonial blend was the hotel and location for the MeSy Rock Mechanics Workshop, New Delhi

1

Stress in Rock Mechanics

Fritz Rummel

ABSTRACT

Rock Mechanics may be defined as the science to understand the response of crustal rocks to acting natural stresses (plate tectonics), or to artificially applied stresses due to large-scale man-made engineering structures. Hydraulic fracturing developed since the mid-sixties has become the most powerful method to determine both, orientation and magnitude of crustal stresses. The fundamentals of the classical hydrofracturing concept and the fracture mechanics approach are outlined, and a case history of crustal stress measurements for the Hong Kong region is presented.

1.1 INTRODUCTION

Rock Mechanics may be defined as the science to study and understand the response of crustal rocks to applied or acting stresses. Crustal rocks contain fractures of various dimensions from the microcrack scale to joints and large faults. At low stresses, the fracture may close and the rock mass may behave elastically. At higher stresses, the fractures may grow or shear and new fractures may originate within intact rock sections. Similarly, crustal rocks react to changes of the fluid pressure. If fluid is injected into boreholes, the fluid will migrate into existing fractures and decrease the normal stresses acting across the fractures such that shear will occur along the fracture plane, or the pressurized fluid will open existing fractures and may also induce fracture growth. The latter process generally is called hydraulic fracturing.

The reaction of crustal rocks with respect to stresses or fluid pressure changes has both, implication on natural processes such as crustal instability (e.g. earthquakes) and on the stability of man-made engineering structures (mining, caverns, tunnels, or dams). The problem may be dramatically interrelated when an earthquake affects a nuclear power project or an underground nuclear waste storage deposit. On the other hand, induced hydraulic fracturing in crustal rocks is used for productivity enhancement in the oil and gas industry or for the creation of an underground heat-exchanger for geothermal energy exploitation from hot crystalline rock. In order to understand the different instability processes and phenomenas, and to optimize hydraulic stimulation techniques the knowledge on crustal stresses is essential.

1.2 CRUSTAL STRESS MEASUREMENTS

Estimates and data on the state of stress in the Earth's crust may be derived from many sources. Until 1970 most stress measurements have been motivated by engineering needs for the design of underground constructions such as tunnels, underground power houses, and mining excavations. Therefore, stress tests were mostly carried out at shallow depth and in close vicinity to the underground openings. Stress measurement techniques developed in the past were particularly designed for such situations. They may be summarized as stress or strain relief methods. A review on the different methods can be found, for example in Amadei and Stephansson (1997).

An alternative method to directly derive stress information from fluid injection tests in boreholes was first proposed by Scheidegger (1962), Fairhurst (1964), and Kehle (1964) on the basis of earlier conclusions derived from hydraulic fracturing laboratory work by Hubbert and Willis (1957). The classical H&W concept can be summarized as follows:

- A hydraulic fracture will be initiated in the rock of a borehole wall when the fluid pressure in the borehole will exceed the minimum tangential stress.
- For homogeneous, isotropic rock the induced fracture will propagate in the direction of least resistance, i.e. perpendicular to the last principal stress.
- As soon as fluid injection is interrupted the shut-in pressure balances the normal stress acting across the fracture plane, i.e. the shut-in pressure is a measure for the least principal stress.
- If, subsequently, the deflated fracture is re-pressurized the fracture will re-open when the minimum tangential stress is again exceeded.

For the case of a vertical borehole (Fig. 1.1) oriented parallel to the intermediate principal far-field stress S_v (vertical stress), the H&W concept can be summarized by the following simple equations:

Fracture generation $\qquad P_c = 3 \cdot S_h - S_H + P_{co}$ (1a)

Re-pressurization $\qquad P_r = 3 \cdot S_h - S_H$ (1b)

Shut-in $\qquad P_{si} = 3 \cdot S_h$ (1c)

Tensile strength $\qquad P_{co} = P_c - P_r$ (1d)

where S_h and S_H are the minor and major horizontal principal stresses, and P_c, P_r, P_{si} are the rock breakdown, the refrac, and the shut-in pressure, respectively (Fig. 1.2). Thus, the determination of the characteristic pressure values during a hydrofrac test directly allows the determination of the horizontal principal stresses.

However, the simple H&W concept is only valid if several additional severe conditions are fulfilled:

- the rock around the borehole must be homogeneous, isotropic, and linear elastic.
- the rock should be impermeable and unfractured so that the pressurizing fluid does not penetrate into the rock and change the stress field and the pore pressure around the borehole prior to hydraulic fracturing.

None of these conditions exist, in particular, crustal rocks per se contain fractures on all scales from intrinsic micro-cracks to the natural joint network and large faults. The possibility to derive stresses from

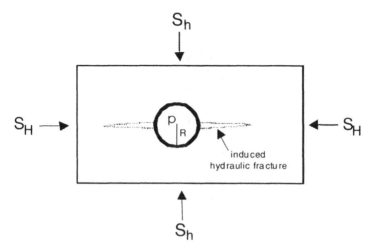

Fig. 1.1 Vertical borehole of radius R subjected to the far-field horizontal principal stresses S_H and S_h, and pressurized by a fluid pressure p to induce a hydraulic fracture

hydraulic injection tests on pre-existing fractures with arbitrary orientation was addressed by various authors in the mid-eighties when the HTPF-and PSI-methods were formulated (see Rummel 2002). Both methods concentrate on measuring precisely the pressure to keep the stimulated fracture open against the normal stress acting across the fracture ($P_{si} = \sigma_n$, Fig. 1.3). Pressure data from test on several fractures with different known orientations then can be used in an inversion calculation to derive the principal stresses. Necessary numerical calculation codes are available (Baumgärtner 1987).

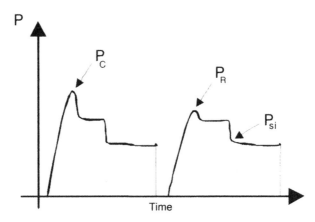

Fig. 1.2 Schematic hydraulic fracturing pressure record with fracture initiation (P_c breakdown pressure), fracture re-opening (P_r refrac pressure), and pressure balance during fracture closure (P_{si} shut-in-pressure)

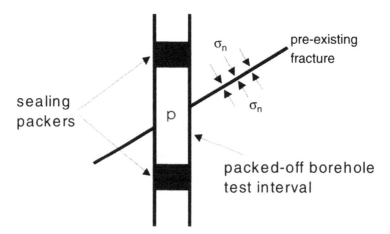

Fig. 1.3 Pressurization of an inclined pre-existing fracture against the acting normal stress σ_n

A new impetus to understand the growth of fracture by fluid injection came from fracture mechanics developed in material science (Anderson, 1991). A rather comprehensive review on fracture mechanics of rock is given by Atkinson (1987). For the growth of hydraulic stimulated fracture we assume a symmetric double crack of half length 'a' extending from a borehole of radius R in an otherwise intact plate subjected to compressive field stresses S_H and S_h (Fig. 1.4). For simplicity, the crack is oriented perpendicular to S_h. Fluid pressure p is applied to the borehole wall, and the fluid may penetrate into the double crack and pressurize the crack by a pressure $p_a(x)$. Inspite of this rather complex loading situation the intensity of the stress field around the crack tips can easily be specified using the principle of superposition of stress intensity factors for each loading source:

$$K_I (S_H, S_h, p, p_a(x)) = K_I (S_H) + K_I (S_h) + K_I (p) + K_I (p_a) \qquad (2)$$

K_I denotes the stress intensity factors for crack opening or mode I fractures. The stress intensity factors for most loading situations can be found in Sih (1973). Substituting into above relation and solving the equation for the critical borehole pressure p_c when crack growth starts leads to

$$p_c = \frac{K_{IC}}{(h_o + h_a) \cdot \sqrt{R}} + k_1 \cdot S_H + k_2 \cdot S_h \qquad (3)$$

where K_{IC} is the critical stress intensity which is called fracture toughness, and h_o, h_a, k_1, and k_2 are normalized dimensionless stress intensity functions which only depend on the crack size 'a' and the specific loading situation. We already see the similarity of eq. (1a) for fracture generation and eq. (3) for crack instability. For zero far-field stresses ($S_H = S_h = 0$) eq. (3) leads to a fracture mechanics physical definition of the macroscopic parameter of hydraulic tensile strength p_{co}:

$$p_{co} = \frac{K_{IC}}{(h_o + h_a) \cdot \sqrt{R}} \qquad (4)$$

The relation clearly indicates that tensile strength is size dependent (R borehole radius) and also on the rate of fluid injection which determines the pressure distribution within the cracks and, therefore, the value of h_a. Further, the stress coefficients of S_h and S_H, k_1 and k_2 may be compared to the values $k_1 = 3$ and $k_2 = -1$ of the classical eq. (3) for zero crack length of an intact (crack free) rock. Thus, in the fracture mechanics approach to hydraulic fracturing the observed breakdown pressure p_c or the strength parameter p_{co} have lost their significance and simply define the instability condition of a crack of a given length during the hydraulic fracturing process in the rock subjected to the far-field stress conditions.

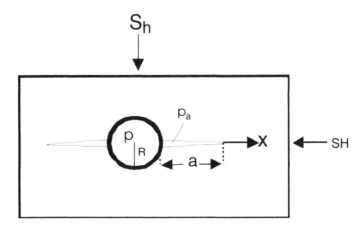

Fig. 1.4 The fracture mechanics approach: A borehole with a pre-existing micro fracture of length a subjected to the far-field stresses S_H and S_h and pressurized by fluid pressure in the borehole and within the fracture, p and p_a

1.3   HYDROFRAC STRESS CASE HISTORY

Several excellent examples for crustal stress measurements could be presented where either hydrofrac tests were conducted to great depth, like in the German 9 km deep borehole of the Continental Deep Drilling Project (Brudy et al., 1997), or hydrofrac tests were carried out in numerous boreholes as part of the site investigation for a large-scale engineering project within a certain region. Such examples, are mostly associated with fast-speed railway tunnels (a good example is the tunnel project between Stuttgart and Munich, Germany (Rummel, 2002), or the trans-alpine tunnel projects) or large mining operations like coal mining in the Ruhr carboniferous (Rummel, 2002) or copper mining in Chile (the author has just conducted hydrofrac stress tests in numerous boreholes to 1000 m depth in the Chuquicamata and Andina region). The major objective in such projects is to determine the regional stress regime as a stability design parameter for planned earth constructions. This is, in particular, important for areas with known plate tectonic activities.

Such an area is Hong Kong in the vicinity of the triple junction of the Philippine, the south eastern Asia, and the Indian plate in the south. A number of large-scale engineering projects are presently in progress, such as mass transport underground railway systems, water reservoirs, and waste storage caverns. Since 1990, about 300 hydrofrac stress measurements were conducted in more than 30 boreholes drilled to a depth of about 200 m, between Tsing Yi Island in the west to Tseung Kwan O in the east (Fig. 1.5). The boreholes were drilled in different geological formations such as sandstones, quartz conglomerates, siltstones and

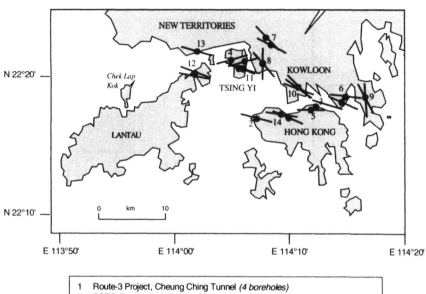

1	Route-3 Project, Cheung Ching Tunnel *(4 boreholes)*
2	SSDS Cavern Facilities, Mt. Davis *(4 boreholes)*
3	SSDS Tunnel Alignment, North Point *(1 borehole)*
4	Lantau Fixed Crossing – Tsing Ma Bridge Anchorage *(2 boreholes)*
5	MTR Quarry Bay Extension *(1 borehole)*
6	MTR Tseung Kwan O Extension, Chiu Keng Wan Shan *(2 boreholes)*
7	Additional Treatment and Water Transfer Facilities, Golden Hill Country Park and Wo Yi Hop Village *(2 boreholes)*
8	KCRC West Rail Central Section Extension, Ha Kwai Chung *(2 boreholes)*
9	MTR Tseung Kwan O Extension, Pak Shing Kok *(2 boreholes)*
10	Central Kowloon Route *(3 boreholes)*
11	Route-9 Project *(2 boreholes)*
12	Penny's Bay Rail Link Project *(2 boreholes)*
13	Route-10 Project, Tsing Lung Tau Section *(5 boreholes)*
14	KCRC Shatin to Central Link Project *(2 boreholes)*

Fig. 1.5 Location of hydrofrac borehole test sites in the Hong Kong area, together with the observed direction of maximum horizontal compression

shales, but mostly in heavily fractured granites, grand diorites, and volcanic rocks. Therefore, the classical H&W approach was not applicable for the analysis of the hydrofrac pressure records (Fig. 1.6). The analysis was conducted by the PSI-method. In addition, due to the pronounced topography in the Hong Kong area, the stress analysis had to be performed in combination with numerical model calculations (e.g. FLAC/Itasca) by assuming rock isotropy, elasticity, gravitational loading, and boundary conditions. The model calculations however, demonstrate that inspite of topographical effects the vertical stress can be taken as a principal stress component for a depth below a few tens of metres.

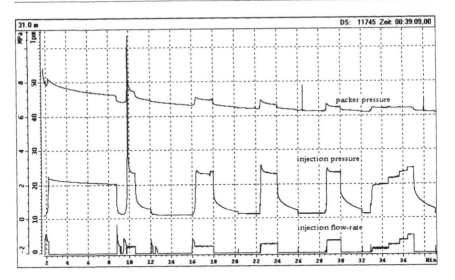

Fig. 1.6 Typical hydrofrac test pressure record for a test section at 31 m depth in borehole no. SAW620/DH/035 at Admiralty, Hong Kong island.

Neglecting the scatter, the hydrofrac tests yield a mean orientation of the maximum stress component S_H of N 108 ± 28 degrees. (Fig. 1.7). This WNW-ESE direction is in agreement with the stress orientation observed for SE China and the direction measured in deep boreholes in the Pearl River Mouth basin, 220 km south of Hong Kong (World Stress Map for SE Asia (1999)). This direction essentially reflects the SE-NW-motion of the Philippine plate (Park, 1988).

The stress magnitudes for rather shallow depth suggest thrust faulting with $S_v < S_h < S_H$, while deeper stress data indicate strike-slip tectonics with $S_h < S_v < S_H$. The stress magnitude versus depth profiles (Fig. 1.8) can be averaged to

$$k_h = \frac{S_h}{S_v} = \frac{72}{z} + 0.66$$

$$k_H = \frac{S_H}{S_v} = \frac{110}{z} + 1.30$$

where the depth z is measured in metres. These relations may be used for a first order stress estimation for both, the design of engineering structures in Hong Kong and for the tectonic characterization for SE Asia. However, inspite of the large quantity of stress information for the specific area, the data scatter is significant, and the construction is more of detailed stress

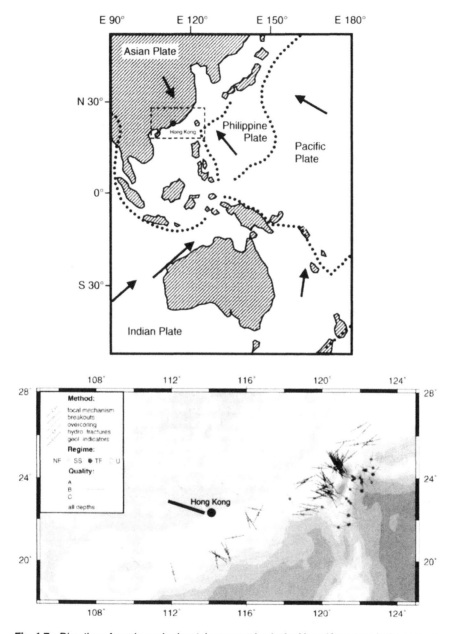

Fig. 1.7 Direction of maximum horizontal compression in the Hong Kong area in the context of plate tectonics

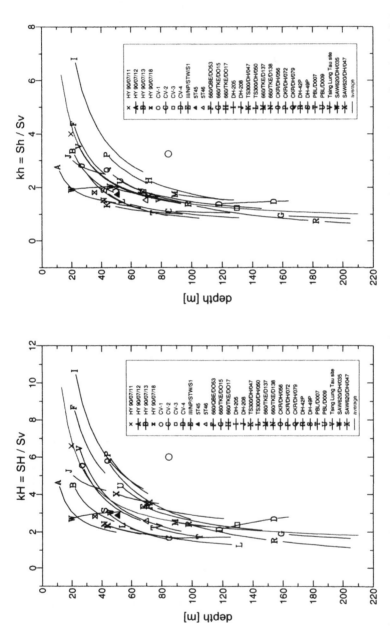

Fig. 1.8 Stress vs. depth profiles measured in boreholes in the Hong Kong area. k_h ration between minor horizontal stress S_h and vertical stress S_v, k_H ratio between major horizontal stress S_H and S_v.

map for an area as small as Hong Kong as it requires a more systematic approach.

REFERENCES

Amadei, B. and Stephansson, O. 1997. *Rock stresses and its measurements.* Chapman & Hall.

Anderson, T. L. 1991. *Fracture mechanics.* CRC Press.

Atkinson, B. K. 1987. *Fracture mechanics of rocks.* Academic Press, Geol. Ser., London.

Baumgärtner, J. 1987. Anwendung des Hydraulic Fracturing Verfahrens für Spannungsmessungen in geklüfteten Gebirge. Doct. Thesis, Berichte Inst. Geophysics, Ruhr-University Bochum, Ser. A, No. 21.

Brudy, M., Zoback, M. D. Fuchs, K. Rummel, F. and Baumgaertner, J. 1997. Estimation of the complex stress tensor to 8 km depth in the KTB scientific drill holes. *JGR, 102:* 18453-18475.

Fairhurst, C. 1964. Measurements of in-situ stresses with particular references to hydraulic fracturing. *Eng. Geol., 2:* 129-147.

Free, M.W., Haley, J. Klee, G. and Rummel, F. 2000. Determination of in-situ stress in jointed rock in Hong Kong using hydraulic fracturing and over-coring methods. *Proc. Conf. Eng. Geol. HK 2000.*

Hubbert, M. K. and Willis D. G. 1957. Mechanics of hydraulic fracturing. *Petrol. Trans. AIME, TP 4597,* 210: 153-166.

Kehle, R. O. 1964. Determination of tectonic stresses through analysis of hydraulic well fracturing. *JGR,* 69: 252-273.

Klee, G., Rummel, F. and Williams A. 1999. Hydraulic fracturing stress measurements in Hong Kong. *Int. J. Rock Mech. Min. Sci.,* 36: 731-741.

Park, R.G. 1988. *Geological structures and moving plates.* Blackie & Sons, Glasgow (e.g. Chapt. 3).

Rummel, F. 2002. Crustal stress derived from fluid injection tests in boreholes. *In: In-situ characterization of rocks* (eds. Sharma and Saxena), chapt. 6: 205-244, Balkema Publishers.

Scheidegger, A. E. 1962. Stresses in the Earth's crust as determined from hydraulic fracturing data. *Geologie und Bauwesen, 27:* 45-53.

Sih, G. C. 1973. *Handbook of stress intensity factors.* Inst. Fracture and Solid Mech., Lehigh Univ., Bethlehem, USA.

2

State of Stress in the Indian Subcontinent

Thirumandas Narayan Gowd

ABSTRACT

National Geophysical Research Institute, in collaboration with Prof. Dr. F. Rummel, Ruhr-University Bochum, Germany, has developed the unique national facility in the country in the year 1985 (Gowd et al. 1986) for measuring in-situ stresses in vertical and near vertical boreholes up to 600 m depth using wireline hydraulic fracturing equipment. In-situ stress measurements were carried out, using this unique facility, at several locations in the country for engineering applications relating to underground mines, power house caverns, as well as for the purpose of scientific studies. Results of the measurements at some of the interesting locations including Hyderabad, Malanjkhand Copper Project, Sardar Sarovar Dam site, Mahabhalipuram south of Chennai, epicentre of the deadliest Latur earthquake at Killari, Rajpura – Dariba mines of Hindustan Zinc Limited are reviewed. The Rock Mechanics group has also determined direction of maximum horizontal principal stress (S_{Hmax}) from the analysis of borehole breakouts in several oil exploration wells as well as from focal mechanisms of several earthquakes in the country. Gowd et al. (1992) prepared stress map of the Indian subcontinent using the S_{Hmax} orientations derived from the hydraulic fracturing stress measurements, borehole breakouts and earthquake focal mechanisms, and recognized four stress provinces within subcontinent. They have interpreted the stress map and concluded that the stress field in the Indian subcontinent is largely determined by the plate tectonic collision processes. Gowd et al. (1996) analysed the mechanics of earthquake activity of the Indian shield as a logical application of the stress map and gained insights into reactivation mechanisms causing intraplate seismicity in the Indian shield. The stress map and its interpretation, and the reactivation mechanisms are also reviewed in this chapter.

2.1 INTRODUCTION

The Indian subcontinent is characterized by high seismicity in the Himalayas and the adjoining northeastern India while the Indian shield was considered until recently as a stable continental region relatively free from earthquakes. However, the Koyna earthquake of December 10, 1967 with a magnitude of 6.7, following the impoundment of water in the Shivaji Sagar Lake in 1962 (Gupta et al., 1969), and a number of other earthquakes of magnitude ≥ 5.0 that occurred within a short period of time (1967 to 1970) in the Indian shield (Chandra, 1977), have given rise to doubts regarding its seismic nature. Following the occurrence of the deadliest earthquake of magnitude 6.4 at Killari near Latur (Maharashtra) on September 30, 1993 (Gupta, 1993), Indian shield is no longer considered as a seismically inactive continental region. The high seismicity of the Himalayan zone and northeastern India, and the moderate seismic activity confined to some linear seismic belts in the Indian shield (Gowd et al., 1996) were attributed to the existence of high tectonic stresses in the region caused by the continental collision between India and Eurasia (Chandra, 1977; Kailasam, 1979; Valdiya, 1989, 1993; Raval, 1993; Khattri, 1994; Gowd et al., 1992, 1996).

A precise knowledge of the state of stress (magnitude as well as direction of tectonic stresses) in the Indian subcontinent is essential to understand the mechanics of plate-boundary earthquakes along the collision zone and intraplate earthquakes in the Indian shield. It also gives insights into plate tectonics and geodynamics of the region. Besides, in-situ stress data is equally important for engineering applications such as designing of underground rock structures (mines, tunnels, etc), planning of large-scale hydraulic fracturing operations for stimulating oil and gas wells, extraction of geothermal energy, disposal of industrial liquid wastes into the subsurface rocks, etc. However, no direct stress measurements were carried out in the country until recently due to lack of expertise. In order to fill this gap, the National Geophysical Research Institute (Hyderabad) has undertaken an Indo-German collaboration project (under the framework of CSIR – KFA agreement) with Prof. Dr. F. Rummel, Ruhr-University Bochum as project leader from Germany and Dr. T.N. Gowd as the project leader from India. They have together developed a unique national facility at NGRI to directly measure in-situ stresses in deep boreholes (upto 600 m depth) of 3-5 inches diameter by employing state-of-the-art wireline hydraulic fracturing technique of Rummel et al. (1983). Under the project, KFA Julich (Germany) donated to NGRI a wireline hydraulic fracturing equipment manufactured by Mesy GmbH (Germany) in the year 1985. The first in-situ stress measurements in India were carried out in the year 1986 at Hyderabad using this equipment (Gowd et al., 1986). A number of scientists of NGRI were trained under this project to gain expertise in wireline hydraulic fracturing technique for evaluating in-situ stresses. Using this unique facility and the expertise acquired under the project, NGRI scientists carried out in-situ stress measurements at several locations in the Indian shield for the purpose of engineering applications and scientific studies. Some of the important results of these measurements will be presented and reviewed in this chapter. Also, stress map of the Indian subcontinent, prepared by Gowd et al. (1992b) using the hydrofrac stress data, maximum compressive stress orientations derived from borehole breakouts of oil wells and earthquake focal mechanisms will be reviewed. Significance of the stress map to the understanding of the plate tectonics and geodynamics of the region will be briefly discussed in the chapter. Mechanisms responsible for the earthquake activity in the Indian shield (intraplate seismicity) have been analysed by Gowd et al. (1996) taking into consideration the stress map (Gowd et al. 1992b). Results of this analysis will be briefly mentioned.

2.2 EXPERIMENTAL TECHNIQUE

2.2.1 Hydraulic Fracturing

The wireline hydraulic fracturing equipment (WHFE) used in our studies is similar to the one developed by Rummel et al. (1983) and was manufactured by Mesy GmbH Bochum. The equipment was described in detail by Gowd et al. (1986). The WHFE comprises a double straddle packer assembly, winch, and a duplex plunger pump (high pressure injection pump) with a heavy duty tripod, 500 m long seven conductor electromechanical cable, and a high pressure rubber hose or stainless steel tube mounted on a four-wheel trailor (Fig. 2.1).

The equipment also includes an electronic data recording unit and an impression tool. Another novel feature of the equipment is its push-pull valve, which is mounted on the straddle packer assembly and is activated

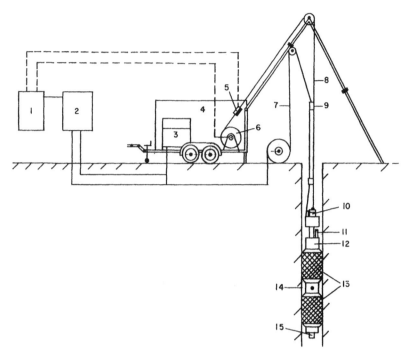

Fig. 2.1 Schematic of the wireline hydrofrac equipment. 1. Electronic recording system, 2. Pressure and flow control panel, 3. Pumping unit, 4. Trailor mounted winch and pumping unit, 5. Rotational transducer to indicate depth, 6. Cable drum, 7. High pressure hose, 8. Seven-conductor logging cable, 9. Clamp to connect hose to cable, 10. Pressure transducer, 11. Release valve, 12. Push-pull valve, 13. Packer elements, 14. Fracture interval, 15. Steel mandrel over which the hydrofrac tool (10-13) is assembled (After Gowd et al., 1986).

to connect the pumping unit through a single pressure hose either to the packers or to the testing borehole sections (TBS) by applying or releasing tension in the logging cable respectively. A pressure release valve, connected to the packer assembly, ensures fast and complete deflation of the packers. The linkage between the logging cable and the packer assembly is provided by the cable head. A pressure transducer, housed within the cable head, measures down hole either packer pressure or fluid pressure in the TBS depending on the position of the push-pull valve. The hydraulic winch and the injection pump are driven by a hydraulic power pack energized by a 22 KW diesel engine. The winch incorporates automatic cable-winding arrangement and a depth counter. The injection pump is rated for a maximum discharge rate of 10 lpm and a maximum injection pressure of 350 bars. The winch and the injection pump are operated through a remote control unit. Strike and dip of the fractures induced or reopened in TBS are obtained by using an impression packer in conjunction with a single shot borehole orientation tool. The electronic recording unit comprises surface and down hole pressure transducers, flow meter, stripchart recorder, and a magnetic tape recorder and facilitates recording of pressure versus time data during hydraulic permeability and hydraulic fracturing tests. Intact sections of a borehole for testing are selected after critically examining the cores recovered from the borehole. The packer assembly is tripped to the deepest TBS and then the packers are inflated to seal it. Then the TBS is instantaneously pressurized to 25 or 50 bars for determining rock permeability. Following this, hydrofrac tests are conducted to initiate a fresh fracture or reopen a pre-existing fracture on the walls of the TBS and propagate it away from the TBS, at full pumping rate of 10 lpm in several phases with 10 to 20 litres of water being injected during each phase. Pressure versus time record is obtained during each frac test (Fig. 2.2). Pressure-time record yields hydrofrac data, i.e., breakdown pressure (P_c), reopening pressure (P_r) and the shut-in pressure (P_{si}). This test procedure is repeated after the packer assembly is moved upwards to the next TBS. After completing all the frac tests, impression of each induced fracture is obtained by using the impression tool. Strike (θ_i) and dip (d_i) of the induced fresh fracs and reopened pre-existing fracs are obtained from their impressions. Hydrofrac data (P_{si}, θ_i, d_i) of induced fresh fracs as well as reopened pre-existing fracs is inverted, by using the shut-in pressure inversion method – SPIM (Cornet and Valette, 1984; Baumgaertner et al., 1986; Srirama Rao and Gowd, 1991; Gowd et al., 1992a; Srirama Rao et al., 1999), for evaluating magnitude of principal horizontal stresses $(S_{Hmax}$ and $S_{hmin})$ and direction of S_{Hmax} (θ^+).

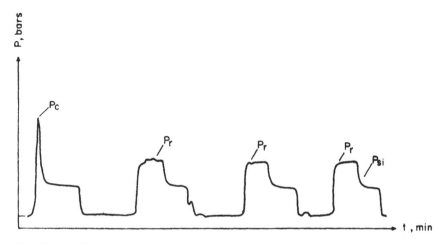

Fig. 2.2 A typical pressure-time record of a hydraulic fracturing test.

2.2.2   Borehole Breakouts and Earthquake Focal Mechanisms

Borehole breakouts and P-axis of earthquakes are the other important stress indicators for deriving S_{Hmax} orientations needed for preparing stress map.

2.3   RESULTS AND DISCUSSION

2.3.1   Hydraulic Fracturing Stress Measurements

In-situ stress measurements were carried out by the NGRI scientists at several locations in different parts of the country using the WHFE donated by KFA Julich (Germany). Results of some of the measurements are presented below.

2.3.1.1   *Stress measurements at Hyderabad (AP)*

The first in-situ stress measurements in India were carried out in the year 1986 in the campus of the National Geophysical Research Institute at Hyderabad (Gowd et al., 1986). A borehole of three- inches diameter was specially drilled to a depth of 176 m for this purpose. Hydrofrac tests were conducted at 12 TBS in the depth range of 43-149 m. Impression tests were successfully carried out at all these 12 TBS. Frac impression data revealed that vertical fractures were induced at eight TBS while pre-existing

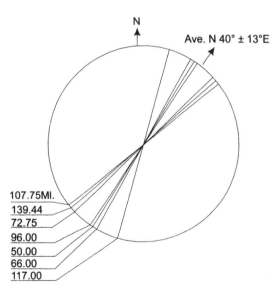

Fig. 2.3 Orientation of the vertical fractures induced during the hydrofrac tests at various depths in the NGRI campus borehole, Hyderabad (After Gowd et al., 1986)

inclined fractures were reopened at remaining four TBS. The azimuth of the induced vertical fractures at all the eight TBS (Fig. 2.3) demonstrated that the induced vertical fractures are consistently oriented in NE-SW quadrant throughout the 50-150 m depth section.

The mean of all these eight orientations was found to be N 40° ± 13°E and indicated, according to the classical method of hydrofrac data interpretation (Hubbert and Willis, 1957), that S_{Hmax} is oriented in N 40°E at Hyderabad. The principal horizontal stresses were also evaluated using the shut-in pressure inversion method (SPIM) mentioned above. The results revealed that the principal horizontal stresses (S_{Hmax}, S_{hmin}) vary with depth as per the following equations:

$$S_{Hmax}(MPa) = 2.7 + 0.045\ Z\ (m)$$

$$S_{hmin}(MPa) = 2.6 + 0.025\ Z\ (m)$$

The plots of S_{Hmax}, S_{hmin} and S_V versus depth (Fig. 2.4) revealed that both the principal horizontal stresses are higher than the vertical stress,

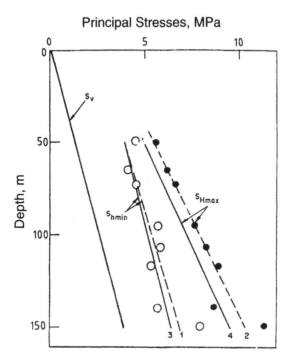

Fig. 2.4 Maximum and minimum principal horizontal stresses (S_{Hmax} and S_{hmin}) and the vertical stress (S_v) as a function of depth (Z) at Hyderabad. 1, 2 are the best line fits to the values (O,●) computed using the classical interpretation method; 3, 4 are the values evaluated using SPIM (After Gowd et al., 1986).

indicating that the former are of tectonic origin. Inversion of the hydrofrac data also yielded the direction of the maximum horizontal compressive stress (S_{Hmax}) as N 35°E, which closely agrees with S_{Hmax} orientation of N 40°E derived from the classical method of hydrofrac data interpretation.

2.3.1.2 *Stress measurements at Malanjkhand (MP)*

Gowd et al. (1986) carried out in-situ stress measurements at Malanjkhand in the year 1986 in one of the exploratory boreholes (Borehole MDN-4) drilled quite far away then from the open-pit mine of the Malanjkhand Copper Project (MCP), see Fig. 2.5.

The MCP area is occupied by granites and a prominent arcuate shaped highly fractured quartz reef. Detailed analysis of the cores recovered from MDN-4 indicated that the top 60 m is the zone of oxidation, and granite, aplite, and metadolerite predominantly occur between 60 m and 400 m depths. Hydrofrac tests were successfully conducted at 15 TBS and frac

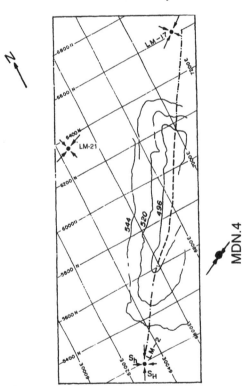

Fig. 2.5 Location map of the four boreholes (MDN-4, LM-2, LM-17, LM-21) in relation to the 544 m, 520 m, and 496 m bench marks of the present open-pit position at the Malanjkhand Copper Project site. The maximum and minimum principal stress directions are indicated by arrows.

impressions could be obtained at six TBS only. The hydrofrac data was processed using SPIM. The results revealed that the maximum horizontal compressive stress is oriented in N 73°E direction. The results showed that the S_{Hmax} and S_{hmin} vary with depth as per the following equations:

$$S_{Hmax} \text{ (MPa)} = 7.8 + 0.037 \; Z \text{ (m)}$$

$$S_{hmin} \text{ (MPa)} = 2.9 + 0.016 \; Z \text{ (m)}$$

These findings indicated that the stress gradients at Malanjkhand i.e. 0.037 and 0.016 MPa/m are lower than those (i.e. 0.045 and 0.025 MPa/m) observed at Hyderabad by Gowd et al. (1986). Later, Srirama Rao et al. (1991) carried out in-situ stress measurements in the year 1991 in three boreholes (LM-2, LM-17, and LM-21) at the open-pit mine of MCP (Fig. 2.5). The results revealed that S_{Hmax} is oriented in N 70°E direction at borehole LM-21. The principal horizontal stresses were found to vary with depth as shown below:

$$S_{Hmax} \text{ (MPa)} = 6.3 + 0.033 \; Z \text{ (m)}$$

$$S_{hmin} \text{ (MPa)} = 5.7 + 0.016 \; Z \text{ (m)}$$

These results indicated that the stress field at LM-21 is similar to the ambient stress field as evaluated at the borehole MDN-4, suggesting that the stress field at the former was not disturbed due to minning as it was located 500 m away from the central axis of the open-pit mine.

The stress field at the other two boreholes i.e. LM-2 and LM-17 was found to be as below:

LM-2:

$$S_{Hmax} \text{ (MPa)} = 8.0 + 0.051 \; Z \text{ (m)}$$

$$S_{hmin} \text{ (MPa)} = 0.5 + 0.029 \; Z \text{ (m)}$$

$$S_{Hmax} \text{ orientation } \theta^+ = N \; 25°E$$

LM-17:

$$S_{Hmax} \text{ (MPa)} = 5.1 + 0.027 \; Z \text{ (m)}$$

$$S_{hmin} \text{ (MPa)} = 4.5 + 0.018 \; Z \text{ (m)}$$

$$\theta^+ = N \; 50°E$$

These results revealed that the stress field at LM-2 and LM-17 was disturbed due to mining as these boreholes were located closer to the mining zone and its central axis. The stress perturbance was more at LM-2 than at LM-17 as the former was lying very close to the open-pit mine. The results also showed that S_{Hmax} orientation rotated anticlockwise towards the central axis of the mining zone. The rotation was more at LM-2 (50°)

than at LM-17 (25°) suggesting that the rotation of the stress field was induced by the mining excavation.

2.3.1.3 Stress measurements at Sardar Sarovar project dam near Vadodara, Gujarat

Gowd et al. (1992a) carried out in-situ stress measurements in the year 1991 at the site of the underground power house cavern located down stream of the Sardar Sarovar dam's right bank (Fig. 2.6). The stress measurements were intended to generate necessary stress data for the purpose of designing a suitable rock – support system to the huge cavern (L = 212 m, W = 23 m, H = 57 m) in order to ensure its stability for safely housing six reversible francis turbine units, each of 200 MW capacity. Two boreholes namely AB-579 and AB-580 were drilled specially for the purpose of stress measurements. AB-579 is located near the cavern, while AB-580 is close to the pressure shafts 5 and 6 (Fig. 2.6). The area of investigation is occupied by basaltic lava flows intruded by dolerite dykes.

Hydrofrac tests were conducted in the depth range of 89 RL to –14 RL in AB-579 and in the depth range of 30.7 RL to –26.3 RL in AB-580.

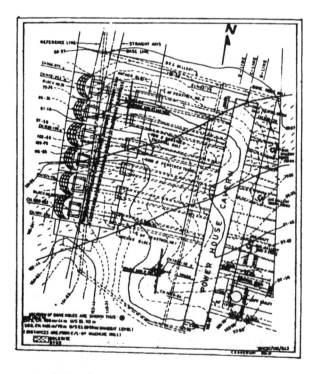

Fig. 2.6 Layout of the power house complex: Location of boreholes AB-579 and AB-580 are shown in the figure (After Gowd et al., 1992a).

Horizontal fractures were reopened at three TBS and vertical/inclined fractures were induced/reopened at four TBS in AB-579, while horizontal fractures were reopened at 4 TBS in AB-580. From the pressure-time records and fracture impressions, hydrofrac data set (P_{SI}, θ_i, d_i) was derived for both the boreholes.

Vertical stress S_V was derived from the shut-in pressure data of the reopened pre-existing horizontal fractures ($S_V = P_{SI}$) for three TBS in AB-579 and for two TBS in AB-580. True depth (Z^*) of these TBS measured from the collar of the respective boreholes were determined. From a plot of S_V Vs Z^*, vertical stress gradient was found to be 0.043 MPa/m, which is absurd as the density of local rocks is about 2.8 gm/cm^3. Since the boreholes were located in a hilly terrain, depth to the TBS in both the boreholes were determined with respect to a common reference plane at 112 RL, i.e., from the collar of AB-579 (Z). A plot of S_V Vs Z (Fig. 2.7) revealed that S_V is charcterized with a gradient of 0.026 MPa/m which

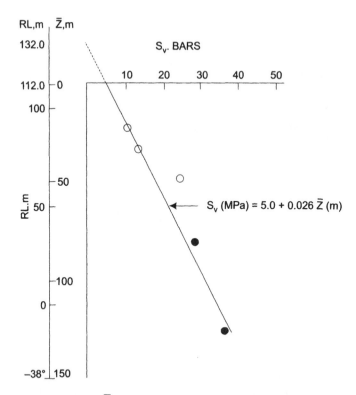

Fig. 2.7 A plot of S_v vs RL, $\overline{Z}$; RL is the elevation of the given site in the area with respect to a common datum plane. Z is the depth of TBS in boreholes AB-579, AB-580 with respect to a common reference plane at 112 RL (After Gowd et al., 1992a).

does not differ significantly from the overburden pressure gradient of 0.028 MPa/m. It was found from this plot that S_V in this area is related to depth with respect to the common reference plane at 132 RL and not to the depth measured from the collar of the respective boreholes.

The hydrofrac data of each of the boreholes (AB-579 and AB-580) was processed separately by using SPIM. The results revealed that the magnitude and direction of the principal horizontal stresses are as below:

Borehole AB-579

$$S_{Hmax} \ (MPa) = 0.85 + 0.07 \ Z \ (m)$$

$$S_{hmin} \ (MPa) = 0.25 + 0.032 \ Z \ (m)$$

$$S_{Hmax} \ \text{orientation} = N \pm 5°$$

Borehole AB-580

$$S_{Hmax} \ (MPa) = 0.81 + 0.065 \ Z \ (m)$$

$$S_{hmin} \ (MPa) = 0.42 + 0.051 \ Z \ (m)$$

$$S_{Hmax} \ \text{orientation} = \text{North} \pm 5°$$

The machine hall, which was under construction then, is located underneath an elevated flat ground. Borehole AB-579 is located 44 m away from the central axis of the machine hall and lies fairly away from the penstocks. As the borehole AB-579 is located at a distance twice the width of the machine hall and lies fairly away from the penstocks, stress field at this borehole might not have been disturbed due to excavation of these structures. In view of this, Gowd et al. (1992a) interpreted that stress field at AB-579 is the ambient stress field prevailing at the site of the machine hall before its excavation, and not the one at AB-580 where the ground was significantly disturbed due to excavation of the penstocks and hence they recommended that the stress field at AB-579 only should be taken into consideration for designing rock support system to the machine hall.

Gowd et al. (1992a) also computed the K-values

$$(K_{Hmax} = \frac{S_{H\,max}}{S_v}), \ K_{hmin} = \frac{S_{hmin}}{S_v}, \ K = \left(\frac{S_{Hmax} + S_{hmin}}{2S_v} \right)$$

as a function of depth for AB-579 (Fig. 2.8) and found that K_{Hmax} increases from 2.9 to 3.5 while K_{hmin} increases from 1.25 to 1.45 in the depth range (–18 RL to 92 RL). They have also recommended that K_{hmin} values only should be considered while designing rock-support system to the machine hall as S_{hmin} is acting perpendicular to its upstream and downstream walls and not K_{Hmax} values as S_{Hmax} is aligned parallel to the machine hall.

STRESS RATIOS

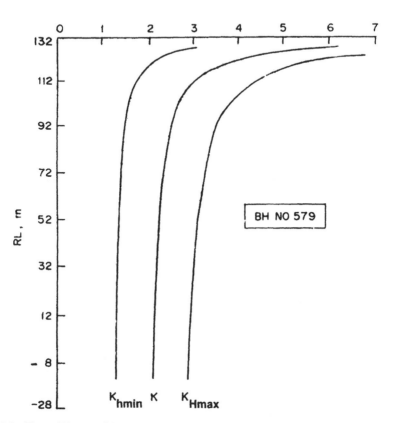

Fig. 2.8 Plots of K_{Hmax} vs RL, K_{hmin} vs RL, K vs RL for borehole AB-579; $K_{Hmax} = (S_{Hmax}/S_v)$; $K_{hmin} = (S_{hmin}/S_v)$; $K = (S_{Hmax} + S_{hmin})/2S_v$; RL is the elevation of a given site in the area with respect to a common datum plane (After Gowd et al., 1992a)

2.3.1.4 *Stress measurements south of Chennai near Mahabalipuram, Tamil Nadu*

In-situ stresses were determined to a depth of 200 m near Mahabalipuram by carrying out hydrofrac tests in two boreholes K-1 and K-2 by Srirama Rao et al. (1997). Hydrofrac data sets (P_{SI}, θ_i, d_i) could be obtained for four TBS in K-1 and for five TBS in K-2. Hydrofrac data sets of the two boreholes were processed separately for deriving in-situ stress field by using SPIM. The results revealed that the principal horizontal stresses at these boreholes vary with depth as shown below:

Borehole K-1

$$S_{Hmax} (MPa) = 4.4 + 0.032 \, Z \, (m)$$
$$S_{hmin} (MPa) = 2.5 + 0.018 \, Z \, (m)$$
$$\theta^+ = N \, 29°E$$

Borehole K-2

$$S_{Hmax} (MPa) = 4.6 + 0.035 \, Z \, (m)$$
$$S_{hmin} (MPa) = 2.5 + 0.016 \, Z \, (m)$$
$$\theta^+ = N \, 32°E$$

From the combined analysis of the hydrofrac data of both the boreholes, stress field near Mahabalipuram was found by them as shown below:

$$S_{Hmax} (MPa) = 4.6 + 0.033 \, Z \, (m)$$
$$S_{hmin} (MPa) = 2.5 + 0.017 \, Z \, (m)$$
$$\theta^+ = N \, 30° \pm 2°E$$

These results showed that the gradients of the principal horizontal stresses at Mahabhalipuram are slightly lower than those at Hyderabad (Gowd et al., 1986). The orientation of S_{Hmax} agrees well with the one at Hyderabad i.e., N 35°E.

2.3.1.5   Stress measurements at Sindesar Khurd near Rajpura— Dariba mines of Hindustan Zinc Limited, Rajasthan

In-situ stress measurements were carried out by Srirama Rao et al. (1999) in a 400 m deep NX size borehole at the site of the proposed shaft at Sindesar Khurd about 6 km away from the Rajpura – Dariba mines near Udaipur (Rajasthan). Calc-biotite schists, Calc-quartzite dolomites, graphite mica schists, calc-silicate bearing dolomites, and quartz veins occur in the area. Hydrofrac tests were conducted at six TBS in the depth range of 60-373 m. The TBS covered all the three types of mica schists namely graphite mica schist at Z = 107 m, carbonaceous mica schist at Z = 160 m, 188 m, and quartz biotite schist at Z = 280, 319, 353 m. The hydrofrac data of the six TBS was processed by using SPIM. The results revealed that the in-situ stress field at Sindesar Khurd is as shown below:

$$S_{Hmax} (MPa) = 1.29 + 0.025 \, Z \, (m)$$
$$S_{hmin} (MPa) = 1.0 + 0.01 \, Z \, (m)$$
$$\theta^+ = N \, 23°E$$

These results showed that Sindesar Khurd is characterized by low horizontal stress gradients of 0.025 and 0.01 MPa/m. These gradients are less than those observed at Hyderabad (Gowd et al., 1986) and at Malanj Khand (Gowd et al., 1986), suggesting that the area is characterized by low horizontal compressive stresses. From the analysis of the results, it was found by them that S_V can be higher than both the principal horizontal stresses from 850 m downwards, implying that the region is subjected to extension tectonics. This could possibly explain, according to them, why several neotectonic extension features such as Ajmer – Junta graben, Masuda – Gulabpur graben, and Karera – Lakhola graben striking in WNW – ESE direction and Chittoor – Nathdwara graben striking in E-W direction are active even today.

2.3.1.6 Stress measurements at the epicentre of the deadliest Latur earthquake of 1993 near Killari (Maharashtra)

In-situ stress measurements were carried out at the epicentre of the Latur earthquake of September 30, 1993 by Srirama Rao et al. (1999) in a 617 m deep BX size borehole (Fig. 2.9) specially drilled for the purpose of

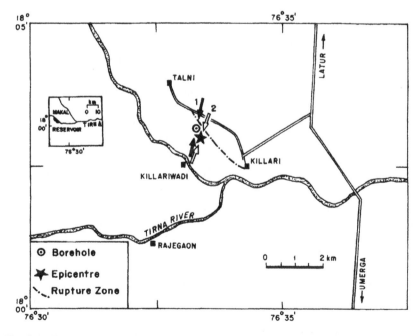

Fig. 2.9 Site map showing the surface rupture, epicentre of the 1993 Latur earthquake, location of the borehole and S_{Hmax} orientation; Epicentre coordinates (18°03'N:76°33'36"E) are the average of the coordinates published by USGS, GEOSCOPE, NGRI, and HRV.1: S_{Hmax} orientation (N 14°E) from the hydrofrac stress measurements; 2: S_{Hmax} orientation (N 22°E) from the fault plane solution by Gupta et al., 1995 (After Srirama Rao et al., 1999).

research jointly by the National Geophysical Research Institute and the Atomic Minerals Division (Gupta and Dwivedy, 1996). The borehole extended through Deccan Traps (0-338 m), intra-trappeans (338-346 m), and the basement rocks of granite and granitic gneisses upto the well bottom (346-617 m). The borehole was cased upto 352 m depth and the open-hole section with a diameter of 60 mm (BX size) extended to the well bottom.

In-situ measurements could be conducted in the shallowest TBS at 373 m and the deepest TBS at 592 m depth. The stress measurements carried out at 592 m depth in the borehole are the deepest of all such measurements made so far in the Indian subcontinent. Due to constraints imposed by the small diameter of the BX borehole on the free movement of the hydrofrac tool, tests could not be carried out by them in the other selected TBS. However, it is for the first time in the world that stress measurements were carried out in a borehole of a small diameter down to a depth of about 600 m. Fracture impression tests revealed that vertical fractures were induced at 373 and 592 m depths. The maximum and minimum principal horizontal stresses (S_{Hmax} and S_{hmin} respectively) were derived from the hydrofrac data using the classical method of hydrofrac data interpretation (Hubbert and Willis, 1957). The S_{Hmax} and S_{hmin} were found to be 16.5 MPa and 9.6 MPa at 373 m depth, and 25.0 MPa and 14.1 MPa at 592 m depth, indicating that the vertical gradients of S_{Hmax} and S_{hmin} in the epicentral zone were 39 MPa/km and 21 MPa/km respectively. The principal horizontal stress magnitudes in the focal region of the Latur earthquake were estimated to be 234 MPa and 126 MPa and the vertical stress was estimated to be 156 MPa (Focal depth: 6 km according to Grad et al., 1997). The results also indicated that the focal region stresses have not undergone any significant change following the earthquake. The results also revealed that S_{Hmax} is oriented in the direction of N 14°E in the Latur area and was found to be in good agreement with the P-axis orientation of N 23°E derived by Gupta et al. (1995) from the fault-plane solution of the deadliest earthquake. The results showed that S_{Hmax} is not oriented at right-angles to the Latur thrust fault (Strike: 126°, Dip: 46° (Seeber et al., 1996) but at 68° only, suggesting that the seismic slip of the Latur earthquake was not completely up dip but had a strike-slip component also. Srirama Rao et al. (1999) showed that the Latur thrust fault could be reactivated if pore pressures within the fault zone were as high as 2.2 times the hydrostatic pressure. For the first time, they applied the concept of fault-valve behaviour as a possible mechanism responsible for inducing such high near-lithostatic pore pressures in the Latur thrust fault zone leading to the occurrence of the Latur earthquake. They cited the temporal variations in the pre-, co- and post-seismic activities of the Latur earthquake as a strong evidence in support of their hypothesis and argued that the globally deadliest Latur earthquake occurred due to rupturing of the over-pressured fault segment of the Latur thrust.

2.3.2 Borehole Breakouts Analysis

Zones of elliptical cross-section, persisting over large depth intervals with a consistent orientation, were observed in oil exploration wells and these elongations were finally understood to be due to spalling or breakout of a borewell (Fig. 2.10) in response to the concentration of compressive stress around it (Zoback et al., 1985; Plumb and Cox, 1987). From direct comparisons, in the same borehole, between breakouts orientations and the principal horizontal stress (S_{Hmax}) directions directly determined from the hydrofrac stress measurements, it was established that borehole elongations are perpendicular to S_{Hmax} direction. Gowd and Srirama Rao

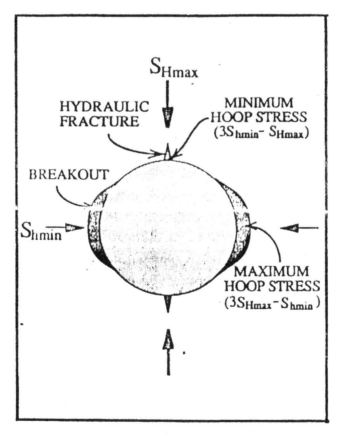

Fig. 2.10 Breakouts in a vertical borehole due to elastic stress concentration around it (borehole) subjected to unequal principal horizontal stresses (S_{Hmax}, S_{hmin})

(1989) analysed wellbore breakouts in several oil exploration wells and determined S_{Hmax} orientation at several sites in the intraplate area of the Indian subcontinent including Jaisalmer basin, Cambay basin, Daman Islands, Bombay offshore, Cauvery basin, Krishna-Godavari basin, Bengal basin, Tripura fold belt, Upper Assam as well as at few locations in the Himalayan Collision zone and Andaman Islands.

2.3.3   P-axis Orientations

P-axis orientations, derived from more than 50 earthquake focal mechanisms, were compiled by Gowd et al. (1992) from the published literature. A sizable number of these solutions were obtained using the Centroid Moment Tensor (CMT) method while the rest were based on the polarities of the first motion. Only those which are characterized by uniform azimuthal distribution of the first-motion readings and constrained by both compressional and dilatational first-motion readings, were included in the compilation. Very few of these compiled P-axis orientations are related to the intraplate area (Indian shield) of the Indian subcontinent while almost all of them are related to the plate boundary zones namely Himalayan collision zone and Indo-Burman subduction zone and indicate stress field in the plate margins only.

2.3.4   Stress Map of the Indian Subcontinent

Gowd et al. (1992b) prepared stress map of the Indian subcontinent showing maximum horizontal compressive stress (S_{Hmax}) orientations derived from the above discussed all the three stress indicators namely hydraulic fracturing stress measurements, borehole breakouts, and P-axis orientations. The map was based on the S_{Hmax} orientation data set of about 100 locations spread over the plate boundary regions and the intraplate area. Only the new stress data sets obtained by the Rock Mechanics Group of NGRI, as discussed above, from the analysis of the wellbore breakouts and hydraulic fracturing stress measurements, made it possible to characterize intraplate stress field in the Indian subcontinent and facilitated the preparation of the stress map of the Indian subcontinent. Four stress provinces have been recognized on the basis of regionally consistent S_{Hmax} orientations. These are the mid continent stress province, the southern shield, the Bengal basin, and the Assam wedge. The mean orientation of S_{Hmax} in the mid-continent stress province, which includes most of the Indian subcontinent except the Bengal basin and the southern shield, was found to be N 23°E (Fig. 2.11) which is subparallel to the direction of compression expected to arise from the net resistive forces at the Himalayan collision zone. Much of southern India (Mysore plateau and the high-grade metamorphic terrain south of the plateau) appear to be

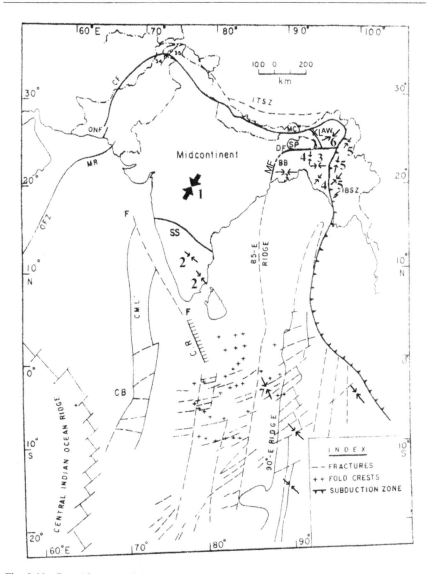

Fig. 2.11 Broad features of the stress map of the Indian subcontinent published by (Gowd et al., 1992). 1: Mean S_{Hmax} orientation in the mid-continent stress province; 2: Typical S_{Hmax} orientation in Southern shield stress province (SS); 3: Typical S_{Hmax} orientation in sedimentary pile of the Bengal basin stress province (BB); 4: Typical S_{Hmax} orientation in the basement and crust beneath the Bengal basin stress province; 5: Typical S_{Hmax} orientation in the subducted slab beneath the Indo-Burman ranges; 6: Typical S_{Hmax} orientation in the intraplate area of Assam wedge stress region (AW); 7: Typical S_{Hmax} orientation in the central Indian Ocean. CB: Chagos Bank; CF: Chaman fault; CMLR: Chagos-Maldiv-Laccadiv Ridge; CR: Comarin Ridge; DF: Dauki Fault; IBSZ: Indo-Burma Subduction Zone; ITSZ: Indus-Tsangpo Suture Zone; KL: Kopili Lineament; MF: Marginal Fault of the Bengal Basin; MCT: Main Central Thrust; MR: Murray Ridge; OFZ: Owen Fracture Zone; ONF: Ornach-Nal Fault; SP: Shillong Plateau.

This sketch is not a political map and does not purport to depict political boundaries. The rough political delineations shown here may not be correct or accurate.

part of a second (Southern shield) stress province characterized by NW oriented S_{Hmax}. The stress field in this province appears close to the intraplate stress field prevailing in the central Indian Ocean. A third stress province was recognized in the Bengal basin which extends from the northeastern margin of the Indian shield to the Indo-Burman ranges. S_{Hmax} within the sedimentary pile of the basin is oriented in E-W direction while P-axis of earthquakes within the basement and crust beneath the basin and within the subducted slab of the Indian plate beneath the Indo-Burman ranges generally trend N-NNE. Interpretation of the results revealed that eventhough the sediments in the Bengal basin are being compressed in E-W direction due to the convergence of the Indian and Burmese plates, the stress field in the basement and crust beneath the Bengal basin and in the subducted slab beneath the Indo-Burman ranges is similar to the one prevailing in the mid-continent stress province. From the interpretation of the stress map, Gowd et al. (1992b) concluded that the Indian subcontinent, including the Indian shield, is being compressionally stressed in NNE direction due to the tectonic collision processes, causing higher stresses in the intraplate region to the south of the collision zone also.

2.3.5 Intraplate Earthquake Mechanisms

Gowd et al. (1996) analysed the mechanisms responsible for causing earthquake activity in the stable continental region of the Indian shield, as a logical application of the stress map prepared by Gowd et al. (1992b). In this regard, seismicity map of the Indian shield showing epicentres of $M \geq 4.5$ was prepared, and seven linear seismic belts were identified on the basis of the seismicity distribution pattern (Fig. 2.12). Also seismogenic faults associated with each of these linear belts were identified. According to them, these seismogenic faults developed in an extensional stress ragime 200-80 Ma and have come under a compressional stress regime following the collision of India with Eurasia about 40 Ma and are being reactivated in response to the intraplate stresses. Mechanisms responsible for the reactivation of the seismogenic faults were analysed considering the in-situ stress field, orientation of the seismogenic faults with respect to the local/regional S_{Hmax} orientations and pore pressures within the fault zones. The results indicated that the seismogenic faults of the Kutch and Latur seismic belts can be reactivated in thrust faulting mode at pore pressures of 2.2 to 2.5 times hydrostatic pressure. The author is of the opinion that the seismic activity in the Narmada-Tapi-Son linear seismic belt is also due to reactivation of the seismogenic thrusts at near-lithostatic pore pressures. Further, the author suggests that the fault-valve behaviour is a possible mechanism for inducing such high pore pressures in the

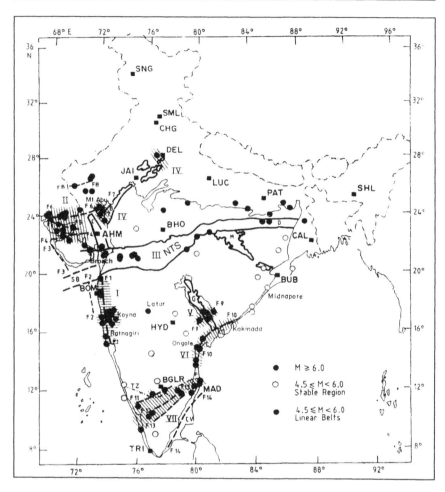

Fig. 2.12: Seismicity map of the Indian shield showing epicentres of earthquakes of magnitude ≥ 4.5. I through VII are the linear seismic belts; I: West Coast seismic belt; II: Kutch seismic belt; III: Narmada-Tapi-Son (Transcontinental) seismic belt; IV: Delhi mobile belt seismic zone; V: Bhadrachalam seismic belt; VI: Ongole seismic belt; VII: Southern granulitic terrain seismic belt. F1: West Coast fault; F2: Off-shore fault; F3: Saurashtra fault; F4: North Kathiawar fault; F5: Kutch fault; F6: Nagar-Parkar fault; F7: Marginal fault of the Delhi mobile belt; F8: A fault in the Jaisalmer area in southern Rajasthan; F9: Bhadrachalam fault, F10 : Marginal fault of the Krishna-Godavari coastal basin, F11 : Kabbani fault, F12 : Bhavani fault, F13 : Attur fault; F14: Boundary fault of the Cauvery basin. B: Delhi mobile belt; NTS: Narmada-Tapi-Son Lineament; M: Mahanadi graben; G: Godavari graben; KG: Krishna-Godavari basin; CV: Cauvery basin; TZ: Transition Zone (After Gowd et al., 1996).

This sketch is not a political map and does not purport to depict political boundaries. The rough political delineations shown here may not be correct or accurate.

seismogenic thrusts of the Kutch and Narmada-Tapi-Son seismic belts, as it was inferred earlier by Srirama Rao et al. (1999) in the case of Latur thrust-fault. Gowd et al. (1996) also showed that the seismogenic faults of the other linear seismic belts namely the West Coast including the Koyna region, Bhadrachalam, Ongole, and Delhi mobile belt seismic zones can be reactivated in strike-slip mode at pore pressure of s 1.3 to 1.7 times the hydrostatic pressures. Recurrence period of the characteristic earthquakes of these seismic belts depends on the rate of build-up of pore pressures to the required critical levels in the fault zones.

ACKNOWLEDGEMENTS

When I joined Prof. Dr. F. Rummel at the Institute for Geophysics, Ruhr-University Bochum, Germany in the year 1971 as a DAAD Research Fellow for my PhD work, it was a matter of great excitement. When Prof. Rummel and I took up the Indo-German collaboration project for measuring in-situ stresses in India, it was a matter of great excitement. Our collaborative research activities have contributed in developing the unique national facility and expertise at NGRI for measuring in-situ stresses in deep boreholes using hydraulic fracturing technique, as well as resulted in generating much needed in-situ stress data leading to preparation of the stress map of the Indian subcontinent. I immensely benefited from my association with Prof. Rummel in specializing myself in Rock Mechanics relating to fracture mechanics and hydraulic fracturing. There cannot be better opportunity than this to express my grateful thanks to Prof. Rummel for all his help and contributions. KFA-Julich Germany donated the wireline hydrofrac equipment to the NGRI in-situ stress project and generously sponsored exchange visits of NGRI scientists to visit Prof. Rummel's laboratory for undergoing training in hydraulic fracturing technique and data processing. I thank KFA-Julich for the same. I am also thankful to my colleagues Dr. M.V.M.S. Rao, Dr. S.V. Srirama Rao, Mr. K.B. Chary, Dr. L.P. Sarma, Mr. N.A. Vijaykumar, Mr. Syed Ali, Mr. C. Prakash and Mr. A. Shyam for their contributions in completing the in-situ stress project of NGRI. I thank the former Directors of NGRI namely Dr. Hari Narain, Prof. V.K. Gaur and Dr. H.K. Gupta for their encouragement and support.

REFERENCES

Baumgärtner, J., Rummel, F., and Chu, Zhaotan. 1986. Wireline hydraulic fracturing stress measurements in the Falkenberg granite massif, In: Falkenberg granite research project (eds. O. Kappelmeyer and F. Rummel), BGR, Hannover.

Chandra, U. 1977. Earthquakes of Peninsular India – A Seismotectonic Study, Bull. Seismol. Soc. Am., 67, 1387-1413.

Gowd, T.N., Srirama Rao, S.V., Chary, K.B., and Rummel, F. 1986. In-situ stress measurements using hydraulic fracturing method, Proc. Indian Acad. Sci. (Earth Planet. Sci.), 95, 311-319.

Gowd, T.N., Srirama Rao, S.V., and Chary, K.B. 1986. Hydraulic fracturing technique for the determination of in-situ stresses at great depth – results of stress measurements at Malanjkhand (MP), Proc. IGU Seminar on "Crustal Dynamics", 125-133.

Gowd, T.N. and Srirama Rao, S.V. 1989. A comprehensive study of wellbore breakouts towards the evaluation of tectonic stress regime in continental India, Tech. Report No. NGRI-89-ENVIRON-59, 93 p.

Gowd, T.N., Srirama Rao, S.V., and Gaur, V.K. 1992. Tectonic stress field in the Indian subcontinent, J. Geophys. Res., 97, 11 879-11 888.

Gowd, T.N., Srirama Rao, S.V., and Chary, K.B. 1992a. In-situ stress measurements by hydraulic fracturing at the underground river bed power house site, Sardar Sarovar Project, Kevadia, Gujarat State, Tech. Report No. NGRI-92-ENVIRON-121, 41 p.

Gowd, T.N., Srirama Rao, S.V., and Chary, K.B. 1996. Stress field and seismicity in the Indian shield: Effects of the collision between India and Eurasia, Pure and Applied Geophysics, 146, No.3/4, 503-531.

Grad, M., Sarkar, D., Duda, S. J., and Kumar, M. R. 1997. Constraints on the focal depth of the Latur earthquake of September 29, 1993, in southern India, Acta. Geophys. Polanica, **XLV,** 93-101.

Gupta, H.K., Narain, H., Rastogi, B.K., and Mohan, I. 1969. A Study of the Koyna Earthquake of December 10, 1967, Bull. Seismol. Soc. Am., 59, 1149-1162.

Gupta, H.K. 1993. The deadly Latur Earthquake, Science, 262, 1666-1667.

Gupta, H. K. et al. 1995. Investigations of Latur earthquake of September 30, 1993, Geol. Surv. India Spl. Pub. No. 27, 17-40.

Gupta, H. K. and Dwivedy, K. K. 1996. Drilling at Latur earthquake region exposes a peninsular gneiss basement. J. Geol. Soc. India, 47, 129-131.

Hubbert, M.K. and Willis, D.G. 1957. Mechanics of hydraulic fracturing, Trans. Am. Inst. Min. Eng. (AIME), 210, 153-168.

Kailasam, L.N. 1979. Plateau uplift in Peninsular India, Tectonophysics, 61, 243-269.

Khattri, K.N. 1994, A hypothesis for the origin of earthquakes of peninsular India, Current Sci., 67, 590-597.

Plumb, R.A. and Cox, J.W. 1987. Stress directions in eastern North America determined to 4.5 km from borehole elongation measurements, J. Geophys. Res., 92, 4805-4816.

Rummel, F., Baumgärtner, J. and Alheid, H.J. 1983. Hydraulic fracturing stress measurements along the eastern boundary of the SW-German block. In: Hydraulic fracturing stress measurements, Nat. Acad. Sci., 3-17 pp (Washington).

Seeber, L., Ekstrom, G., Jain, S. K., Murty, C. V. R., Chandak, N., and Ambruster, J. G. 1996. The 1993 Killari earthquake in central India: A new fault in Mesozoic basalt flows? J. Geophys. Res., 101, 8543-8560.

Srirama Rao, S.V., Chary, K.B., and Gowd, T.N. 1991. Tectonic stress field at the mines of Malanjkhand Copper Project (Hindustan Copper Ltd.), Madhya Pradesh, India, Tech. Report No. NGRI-91-ENVIRON-110, p 25.

Srirama Rao, S.V., Chary, K.B., and Gowd, T.N. 1997. In-situ permeability and tectonic stress field at Mahabhalipuram, Tamil Nadu, Tech. Report No. NGRI-97-Litho-208, p 34.

Srirama Rao, S.V., Chary, K.B., and Gowd, T. N. 1999. Ambient stress at the Malanjkhand Copper Mine Project, Madhya Pradesh, India, J. of Rock Mech. & Tunnelling Tech., 5(1), 35-46.

Srirama Rao, S.V., Chary, K.B., Gowd, T.N., and Rummel, F. 1999. Tectonic stress field in the epicentral zone of the Latur earthquake of 1993, Proc. Indian Acad. Sci. (Earth Planet. Sci.), 108 (2), 93-98.

Srirama Rao, S.V., Chary, K.B., Syed Ali, Vijay Kumar, N.A., Somvanshi, V.K., and Gowd, T.N. 1999. In-situ permeability of the subsurface rocks and tectonic stress field at Sindesar Khurd near Rajpura-Dariba Mines of Hindustran Zinc Ltd., Rajasthan, Tech. Report No. NGRI-99-Litho-251, p 35.

Valdiya, K.S. 1989. Neotectonic Implication of collision of Indian and Asian Plates, Indian Journal of Geology, 61, 1-13.

Valdiya, K.S. 1993. Latur Earthquake of 30 September 1993: Implications and Plannings for Hazard-preparedness, Current Science, 65, 515-517.

Zoback, M.D., Moos, D., Mastin, L., and Anderson, R. N. 1985. Wellbore breakouts and in-situ stress, J. Geophys. Res., 90, 5523-5530.

3

Hydraulic Fracturing Stress Measurements from Underground Openings in India—A Few Case Studies

Amalendu Sinha

ABSTRACT

Rapid advancement in numerical modelling and computational techniques has made Rock Mechanics a practical tool for the planners, designers, and researchers. The wide acceptance of this subject is mainly due to better approaches for understanding the physical behaviour of rock mass and evaluation of stress regime. Central Mining Research Institute (CMRI), since long is conducting various research programs for developing state-of-the-art knowledge base on these subjects. Evaluation of pre-excavation in-situ stress fields was done, by using hydraulic fracturing technique, in different underground hard rock openings in mines, tunnels, and caverns. In Zawar and Rajbura Dariba group of lead-zinc mines (Rajasthan, India), measurements were conducted from underground. Different generations of folds and faults were present in these base metal deposits. In Mochia, Balaria, and Zawarmala mines of Zawar area, the depth of measurements varied from 95 m to 500 m and magnitude of major horizontal stress varied from 8.2 MPa to 28.6 MPa. It was observed that directions of major horizontal stress in these three mines had relation with the major geological structures present in this area. In Rajbura Dariba mine the major horizontal stress was 27.0 MPa at around 200 m depth and 28.4 to 28.8 MPa at around 250 m depth and the direction varied from N280„a to N290o. The minor horizontal stress, at these depths, varied from 15.2 MPa to 15.5 MPa. Similar studies were also conducted in a uranium mine, located in Singhbhum Thrust Belt (Jharkhand, India), from two underground openings. The magnitudes of major horizontal stress at 75 m and 140 m depths were 7.4 MPa and 9.8 MPa respectively towards NNW. Measurements were conducted in two hydro-electric projects in the state of Maharashtra (India) for the design of underground power houses in the volcanic rock (basalt) of Decan Trap. Compared to the hard rock mining areas these locations were found to have low stress regime. The magnitude of major horizontal stress varied from 7.6 MPa to 4.4 MPa due NE, within a depth range of 70 m to 263 m. The stress values determined in underground mines and tunnels/caverns were used to evaluate the stability of openings mainly using numerical modelling techniques.

3.1 INTRODUCTION

Underground openings in rock involve application of the science of Rock Mechanics for design analysis. Rapid advancement in numerical modelling and computational techniques has made Rock Mechanics a practical tool for the planners, designers, and researchers. The wide acceptance of this subject is mainly due to better approaches for understanding the physical behaviour of rock mass such as rock mass classification systems and in-situ stress measurements. Because they are the precursors for any in-depth Rock Mechanics study controlling realistic output and rational design analysis.

Central Mining Research Institute (CMRI), since long is conducting various research programs for developing state-of-the-art knowledge base on these subjects. First systematic in-situ stress measurements by CMRI were conducted in 1988 using hydraulic fracturing technique in Rajpura Dariba lead-zinc mine in Rajasthan. This study was carried out in collaboration with National Geophysical Research Institute (NGRI), Hyderabad. Since then CMRI has conducted quite a few number of studies in different underground hard rock openings in mines, tunnels, and caverns. This chapter embodies, in brief, the outcome of these studies and applications of stress measurement results in stability analysis.

3.2 TECHNIQUE AND EQUIPMENT

The conventional hydraulic fracturing technique was adopted for stress measurements at all the experimental sites. Tensile fractures were induced and breakdown pressure (P_c), reopening pressure (P_R), and shut-in pressure (P_{si}) were recorded. The impressions of the fractures were obtained on an impression packer wherefrom their orientations were determined.

The experiments were conducted both in horizontal and vertical holes. The test locations in each hole were selected after studying the borehole core samples to avoid fractured zones. At each location, prior to pressurizing for inducing fracture, permeability tests were performed and generally further experiments were not conducted if highly permeable zones were encountered. If the experiments were satisfactory, the fracture impressions were obtained. Where the axial fracture was induced in a vertical hole, the magnitudes of major horizontal stress (S_H) and minor horizontal stress (S_h) were calculated based on classical approach using the experimental values P_c, P_R, and P_{si}. The directions of these stresses were determined from the orientation of the axial fracture. The horizontal holes were drilled in the direction of one of the principal horizontal stresses, generally in the direction of S_h. The magnitudes of vertical stress (S_v) and one of the principal horizontal stresses could be determined wherever axial fractures were obtained.

The experiments were conducted with Prefrac 1/2″ System, specially developed for stress measurement from underground galleries and tunnels by MeSy, Germany. With this system, experiments could be conducted in boreholes of 56 mm diameter up to a depth of 30 m. The schematic diagram of this system is shown in Fig. 3.1.

3.3 STRESS MEASUREMENTS IN UNDERGROUND NON-COAL MINES

3.3.1 Rajpura Dariba Mine

Rajpura Dariba is an underground lead-zinc mine of Hindustan Zinc Ltd (HZL). It is located 76 km north-east of Udaipur city in Rajasthan (India). In this mine in-situ stress measurement by hydraulic fracturing technique was undertaken for stability analysis and optimization of stoping parameters. All the measurements were done from the drives in the underground at 200 MRL and 250 MRL (surface is approximately 500 MRL).

The main rock types in the mine area are a) calc biotite schist b) calc quartzite/silicious dolomite c) graphite mica schist d) calc quartzite/calc silicate bearing dolomite, and e) quartz veins. Among these calc quartzite/calc silicate-bearing dolomite is the principal host of sulphide bearing mineralization and occupies the core of Rajpura Dariba synform.

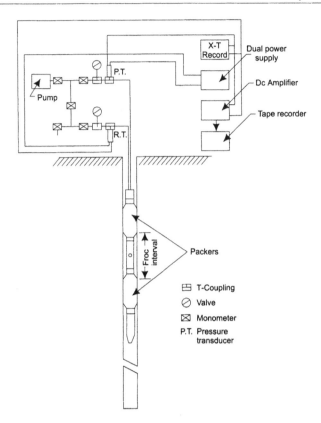

Fig. 3.1 Schematic diagram of hydraulic fracturing equipment

The main lode in this mine extended over 1,700 m strike length and is split into two ore bodies viz. south and north lodes by a barren stretch of 300 m.

Experimental sites

One of the major considerations for selection of sites was to avoid the zones where the existing stoping operation would influence the virgin stress field. Boreholes were drilled in the locations away from stoping areas. In 250 MRL (approx. depth from surface 250 m), experiments were conducted in north lode in one vertical borehole and two horizontal boreholes drilled due N 45°E and S 45°E. All the boreholes were approximately 30 m deep. In the vertical hole many fractured zones were anticipated from the borehole cores and only three locations were found suitable for conducting the experiments. In the horizontal hole drilled due N 45°E six experiments were conducted and in the other borehole five tests were carried out.

In 200 MRL (approx. depth from surface 300 m) experiments were conducted in the boreholes (approx. 30 m) drilled both in north and south lodes. In the north lode experiments were conducted in one vertical borehole and two horizontal boreholes drilled due N 45°E and S 45°E. In the south lode experiments were conducted in one vertical hole and one horizontal hole (direction N-E). In both the horizontal holes drilled in north lode experiments were conducted at four different depths. In the vertical hole at the same location three experiments were conducted.

Results

The hydraulic fracturing experiments at 250 MRL and 200 MRL of Rajpura Dariba Mines of HZL shows:

    i. At 250 MRL:

                Major horizontal stresses: 27.0 MPa at N 280°

                Minor horizontal stresses: 15.2 MPa at N 190°

    ii. At 200 MRL:

      South lode:

                Major horizontal stresses: 28.4 MPa at N 280°

                Minor horizontal stresses: 15.5 MPa at N 190°

      North lode:

                Major horizontal stresses: 28.8 MPa at N 290°

                Minor horizontal stresses: 15.5 MPa at N 200°

   iii. Vertical stress as determined at 200 MRL was 12.2 MPa.

   iv. Overall, the observed direction of horizontal stresses confirm to the minerilization in the isoclinal synform trending in N-S direction.

3.3.2 Zawar Group of Mines

The mines of Zawar group, belonging to Hindustan Zinc Ltd., are located about 43 km south Udaipur city in Rajasthan. In this area stress measurements were conducted in three lead-zinc mines namely, Mochia, Balaria and Zawarmala.

The rocks of the Zawar area are a part of the Aravalli system of middle Cambrian age (Straczek & Srikantan, 1960). The principal rocks in the Zawar area are phyllite, slate, conglomeratic and gritty graywake, dolomite, quartzite, and quartzite conglomerate. The lead-zinc mineralization occurs almost entirely in the dolomite of certain stages.

The rocks generally dip steeply and in places are overturn. They are folded into isoclinal and non-isoclinal folds whose axis have a moderate to steep plunge. Major longitudinal faults disrupt the symmetry of every major fold in the area. The rocks have undergone two major tectonic cycles and each of these cycles lead to development of folds and system of complex faults. The tectonic forces of the first period were directed on an east-west axis and produced folds oriented in a northerly direction. The forces of second period were apparently directed on north-easterly trending axis and refolded earlier folds on a westerly trending axis and further compressed into tight isoclinal structures.

Two major folds viz., Sisa Magra anticline and Zawarmala anticline are present in this area with some other folds. The Mochia and Balaria mines are situated in Sisa Magra anticline whereas the Zawar mine is situated in the Zawarmala anticline as shown in the geological map of the area in Fig. 3.2.

Experimental sites

The hydraulic fracturing experiments were conducted in total eight boreholes, each about 30 m deep. Out of these, experiments in two boreholes were not successful. In Mochia mine, the experiments were conducted at the ninth level (depth 500 m) in one vertical and one E-W trending horizontal boreholes. In both the boreholes, the measurements were done at four different depths.

In Balaria mine, the experiments were conducted at two different levels. At 378 MRL (depth 78 m), experiments were conducted at two different locations in a vertical borehole. At 105 MRL (depth 315 m) the stress measurement was done in one vertical and one horizontal (trend N 200°) boreholes. The tests were conducted at three different locations in each of these boreholes.

In Zawarmala mine, experiments were successful in one vertical borehole drilled from a ramp at 335 MRL (depth 225 m). Experiments in the other two boreholes were not successful in this mine because of numerous fractured zones in the boreholes.

Results

From the experimental data magnitudes of the major horizontal stress (S_H) and the minor horizontal stress (S_h) have been calculated. These are presented in Table 3.1.

The relationships obtained between magnitudes of horizontal stresses and depths are as follows:

Table 3.1   *Magnitude of in-situ stresses at different depths*

Mine	Level from which measured (m)	Total depth from surface (m)	S_H (MPa)	S_h (MPa)
Balaria	378 MRL	95.62	8.2	5.6
Zawarmala	355 MRL	235	17.7	9.6
Balaria	105 MRL	315	20.4	11.8
Mochia	9th Level	500	28.6	—

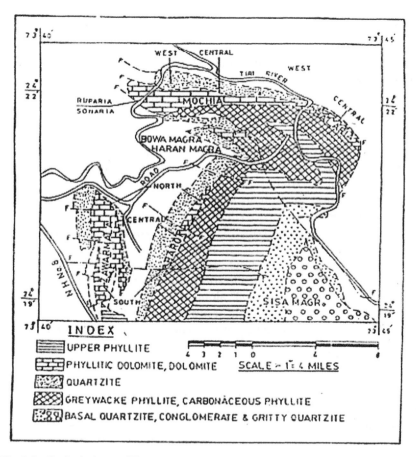

Fig. 3.2   Geological map of Zawar area

$$S_H = 4.6 + 0.049 \times D$$

$$S_h = 3.0 + 0.028 \times D$$

Where, S_H and S_h are expressed in MPa and D is in metre.

In this area, it appears that the directions of major horizontal stress, as measured in Balaria and Zawarmala mine, have a relation with the major geological structures of the area. The direction of major horizontal stress (S_H) in Balaria mine varies from N 110° to N 120°. This direction is nearly parallel to the direction of the plunge of the crossfold present in this area. The major horizontal stress direction is also sub-parallel to the trend of the system of faults present near this area. The direction of S_H in Zawarmala mine varies from N 156° to N 167°. This direction is also, almost, parallel to plunge direction of Zawarmala anticline, which is NNW.

3.3.4 Narwapahar Mine

The Narwapahar mine of Uranium Corporation of India is located about 20 km south of Tatanagar city in the state of Jharkhand (India). In this mine, stress measurements were conducted in the second and fourth levels where the depths of cover were 75 m and 140 m respectively.

Bhola (1968) has discussed the geology of the area in detail; some relevant points have been mentioned here.

The Singhbhum Thrust Belt, of which Narwapahar area forms a part, is covered by meta-sedimentary and meta-volcanic rocks belonging to the Iron ore series of the Archean age, which is subdivided into three stages namely, Chaibasa stage, Iron ore stage, and Dhanjori stage. The main rock types are garnetiferous muscovite-sericite schist, chlorite-sericite schist, chlorite schist, and quartzite. This thrust zone is structurally an anticlinorium of isoclinally folded northerly dipping rocks, with over-thrusting along the southern limb of the geo-anticline caused by tectonic movements directed from north to south. In this area, the foliations strike at N 60° W-S 60° E and dip at 50° along N 30° E, with variation upto 10° both in the direction and the angle of dip. Some cross faults and strike slip faults have been reported to be present. The rocks are well jointed.

Experimental sites

The experiments were conducted in one vertical borehole at the second level, Band no.3, W-16 crosscut (depth 75 m) and in another vertical borehole at the fourth level, W3 H/W (depth 140 m). In the second level, out of the four locations selected, experiment could be successfully conducted only at 28.52 m depth. In other locations, the straddle packer got jammed. In the fourth level, four locations were selected. Among these locations, experiments could be conducted at two locations. In other two locations experiments could not be conducted because of jamming of straddle packer.

Results

The stress measurements were conducted at the second and fourth levels in this mine. The calculated magnitude from experimental data and the directions of major and minor horizontal stresses are as follows:

2nd level

$$S_H = 7.4 \text{ MPa and}$$

$$S_h = 4.4 \text{ MPa}$$

Direction of S_H is N 350° – N 170° thus, the direction of S_h is N 260° – N 80°

4th level

$$S_H = 9.8 \text{ MPa and}$$

$$S_h = 5.6 \text{ MPa}$$

Direction of S_H is N 343° – N 163° thus, the direction of S_h is N 253° – N 73°

3.4 STRESS MEASUREMENTS IN UNDERGROUND POWERHOUSE

CMRI conducted stress measurements in two underground powerhouses in Maharastra, viz., Koyna Hydel Project and Ghatghar Pumped Storage Scheme.

3.4.1 Koyna Hydel Project

The Koyna Hydel Project Stage I Powerhouse belongs to Maharastra Government. It is located about 350 km southeast of Mumbai city. When this underground powerhouse was under construction in-situ stress measurements were conducted in the machine hall, inter adit, and emergency valve tunnel.

The area falls in the Western Ghats and is surrounded by many ridges. In one such ridge machine hall, inter adit, and emergency valve tunnel are located. The Western Ghat comprises volcanic rock and the excavations were being made through basalt flows of Deccan Trap. The main rock types encountered were compact basalt and amygdaloidal basalt. While the compact basalt was well jointed the amygdaloidal basalt had occasional joints. Gupte et al. (1989) reported that no indication of faulting was found in this area. Presence of any major geological disturbance was neither reported nor observed during excavation.

Experimental sites

The experiments were conducted in five boreholes in three locations. At machine hall a drift tunnel was driven at '0' chainage towards N 155°. One vertical and one horizontal holes were drilled from the drift tunnel for conducting the experiments. The direction of the horizontal hole was N 80°. The direction of this hole was decided after knowing the major horizontal stress. In the vertical hole, four locations and in the horizontal hole, three locations were selected for the test. In the inter adit, the experiments were conducted in a drift tunnel at 105 m chainage. One vertical hole and one horizontal hole towards N 80° were drilled for this purpose. In the vertical borehole, experiments were conducted at six locations. In emergency valve tunnel, the tests were conducted in a vertical hole drilled in a drift tunnel at 305 m chainage. Though in this borehole, attempts were made to conduct experiments at eight different locations, axial fracture could not be obtained. Moreover, many locations indicated permeable zone.

Results

Based on the test results obtained, it was attempted to determine the gradient of the virgin stresses in this area. Considering the values of S_H at machine hall (depth 220 m) and inter adit (depth 70 m) 6.3 MPa and 4.4 MPa respectively, the gradient of S_H in MPa in this area should be as follows:

$$S_H = 3.51 + 0.0126 \times D$$

Similarly, the gradient of S_h can be taken as follows:

$$S_h = 2.52 + 0.004 \times D$$

Where, D is depth in metre, and S_H, and S_h are in MPa.

The average direction of S_H and S_h are N 48° and N 138° respectively.

In Machine Hall, where the depth was 220 m, vertical stress (S_v) measured was 4.4 MPa. Therefore, the gradient was 0.02 x D. In Inter Adit, the measured S_v at 70 m depth was 2.2 MPa. Here, the gradient would be 0.031 × D. Such variation in gradient may be due to the difference in density between the amygdaloidal basalt (2,365 kg/m^3) and that of compact basalt (2,932 kg/m^3). Under such condition, S_v will be influenced by the type of basalt overlying the opening. Moreover, the hilly topography of the area will also influence the state of stress and its effect would be more on S_v as it is more sensitive to depth (D) than S_H and S_h in this area.

3.4.2   Ghatghar Pumped Storage Scheme

Ghatghar Pumped Storage Scheme is the first of its kind in Maharastra state being undertaken by the Irrigation Department of the state. It is located about 160 km north east of Mumbai city. Experiments were conducted in the approach tunnel and machine hall.

The geology of this area has been described in detail in the "Engineering Geological Report Regarding Pressure Shaft (Ghatghar Pumped Storage Scheme)" prepared by Dr. S. S. Marathe. Like Koyna Hydel Project, this area also falls in the Western Ghats and is surrounded by ridges. The rocks of the area include different varieties of Deccan Trap basalts viz., compact porphyritic basalt, amygdaloidal basalt, and volcanic breccia with diverse engineering characteristics.

Experimental sites

The experiments were conducted in four vertical holes located at 300 m chainage, 770 m chainage, 1,290 m chainage in the approach tunnel and at 1354 m chainage in the machine hall. Two drift tunnels, one at 300 m chainage and the other at 770 m chainage were driven for conducting the experiments.

In the vertical hole at 300 m chainage, initially, six locations were selected for conducting the tests. Out of the six locations, tests were conducted at four locations. Two locations were permeable, hence discarded. In the vertical hole at 770 m chainage four locations were selected and tests were conducted at all the locations. At 1,290 m chainage, tests were conducted at seven locations.

At 1,354 m chainage in the machine hall, the tests were conducted at two locations in the vertical borehole. In the horizontal borehole, tests were conducted at three locations.

Results

Based on the magnitude of major and minor stresses and depth of test locations (depth from surface + depth of frac test in vertical borehole), the gradients of S_H and S_h have been determined. At 770 m chainage, the values of S_H and S_h are 5.4 and 3.6 MPa respectively at 151.20 m depth and at 300 m chainage, the values of S_H and S_h are 4.8 and 3.2 MPa respectively at 74.40 m depth. In machine hall at 1,354 m chainage, the S_H was 7.6 MPa at 263 m depth. Based on these experimental results, the stress gradient determined for S_H and S_h are as follows:

$$S_H = 3.65 + 0.01467 \times D$$

$$S_h = 2.77 + 0.00549 \times D$$

Where, D is depth in metre, and S_H, and S_h are in MPa

The average direction of S_H and S_h are N 48° and N 138° respectively.

The magnitude of vertical stress S_v is 4.6 MPa at 263 m depth. Hence, the gradient of S_v would be $0.175 \times D$ in machine hall.

3.5 APPLICATION OF STRESS MEASUREMENT RESULTS

All the stress measurement results discussed above were used in stability and design of major underground projects in civil and mining engineering. This was achieved by employing two-dimensional and three-dimensional numerical modelling based on finite element, boundary element, and finite difference (FLAC) software. These studies involve analysis for evaluation of stability support design and monitoring of ground behaviour using suitable rock mechanics instrumentation.

Two examples of numerical modelling based stability analysis have been discussed in brief here under.

3.5.1 Stability Analysis of a Shaft Pillar in an Underground Hard Rock Mine

In Balaria mine the optimum thickness of shaft pillar 120 MRL was to be determined. For this purpose the influence of the stope (3E stope) on the shaft pillar was studied (Prasad, Sinha, and Prasad 2001). The schematic diagram of the mining geometry is shown in Fig. 3.3. Analysis was

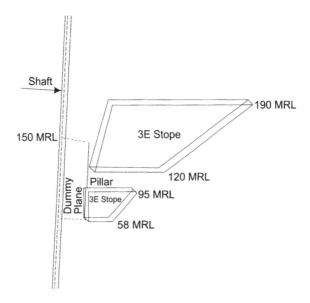

Fig. 3.3 Schematic diagram of shaft and 3E stope with dummy plane for display of results of numerical modelling

conducted using a three-dimensional BEFE package (a hybrid of Boundary Element and Finite Element). The properties of rock and rock mass, which are also major inputs, were determined/estimated from laboratory tests and field geotechnical studies.

After studying various options, it was inferred that a 30 m parting between stope and shaft would be optimum for the stability of the shaft. The contour plot of major and minor principal stresses and safety factor based on rock mass failure criteria (Sheorey, 1997) are shown in Figs. 3.4, 3.5, and 3.6, respectively.

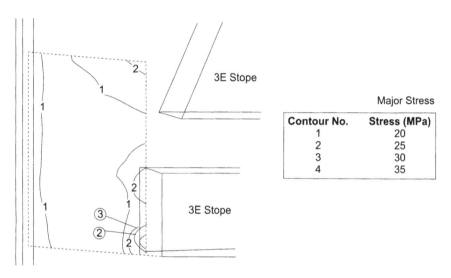

Contour No.	Stress (MPa)
1	20
2	25
3	30
4	35

Fig. 3.4 Contour of maximum principal stress in 30 m shaft pillar

3.5.2 Stability Analysis of Caverns in a Hydroelectric Project

Stability analysis of machine hall and transformer hall caverns at Ghatghar Pumped Storage Scheme in Maharastra state was carried out using a three-dimensional FLAC package (Loui P et al., 2000). Apart from in-situ stress the rock/rock mass properties were also determined from laboratory tests and field geotechnical studies. The three-dimensional isometric view of all the excavations modelled using FLAC 3D are shown in Fig. 3.7. In this study, after determining the principal stresses around the excavations, the safety factor was determined using Sheorey failure criteria (Sheorey, 1997). The safety factor contour along different sections in transformer hall is shown in Fig. 3.8. Similar studies were also conducted for machine hall.

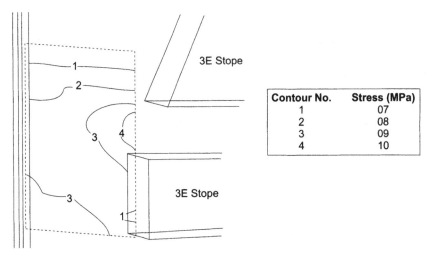

Contour No.	Stress (MPa)
1	07
2	08
3	09
4	10

Fig. 3.5   Contour of minimum principal stress in 30 m shaft pillar

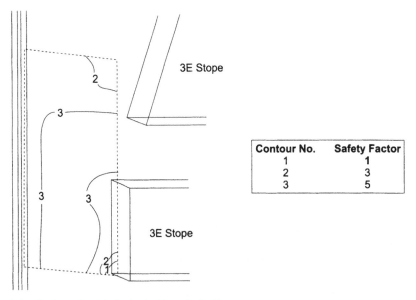

Contour No.	Safety Factor
1	1
2	3
3	5

Fig. 3.6   Contour of safety factor in 30 m shaft pillar

The studies revealed that the excavations are quite stable since none of the elements have safety factor less than 1.0.

3.6   CONCLUDING REMARKS

Till now, CMRI has conducted the measurements by hydraulic fracturing from underground opening in boreholes upto a maximum depth of 30 m.

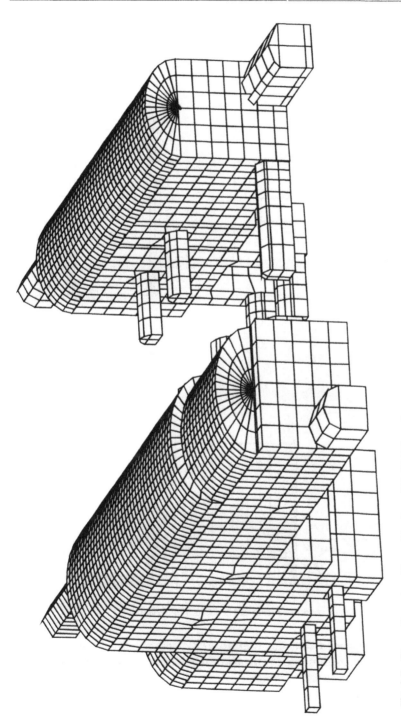

Fig. 3.7   Three-dimensional isometric view of FLAC 3D mode

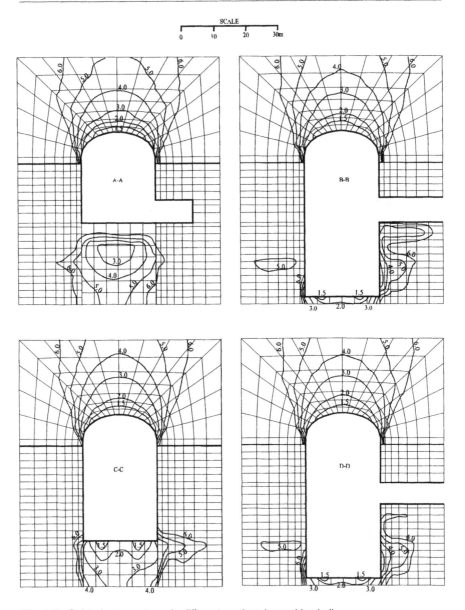

Fig. 3.8 Safety factor contours in different sections in machine hall

It was experienced that the Prefrac 1/2″ System, developed by MeSy, Germany is quite suitable for measurements from underground openings where available space is limited.

The results obtained from hydraulic fracturing experiments have been exhaustively applied for design and stability evaluation by numerical simulation techniques based on FEM, BEM, and FDM in mines and underground caverns under varying geo-environment. The comparison of the outcome of these studies with real behaviour of openings shows that the stress measurement data were very realistic for optimum design analysis.

ACKNOWLEDGEMENTS

The author is indebted to Sri L. Prasad, Dr. M. Prasad and Sri R. Rao who were associated in the hydraulic fracturing stress measurement studies. The author is thankful to Sri A. K. Ghosh, Sri D. G. Rao, and Sri N. Kumar, scientists, CMRI, for their help in preparation of this chapter. The encouragement and guidance provided by Dr. P. R. Sheorey, deputy director, CMRI, Dr. J. L. Jethwa, deputy director, CMRI, and Prof. D. D. Misra, director, CMRI, are gratefully acknowledged.

REFERENCES

Bhola, K.L. 1968. Geology of Narwapahar Uranium Deposit, Int. Geol. Cong. Prauge.

Gupte, R.B., Karmakar, B.M., and Kulkarni, S.R. 1989. Koyna Hydroelectric Project Stage IV, An Engineering Assessment of Geological Features, Internal Report, p 38.

John Loui, P. et al. 2000. Three-dimensional Numerical Modelling of Machine Hall and Transformer Hall Caverns at Ghatghar Pumped Storage Scheme, CMRI Report.

Prasad, M., Sinha, A., and Prasad, L. 2001. Design of Pillars in Hard Rock Mines by Numerical Simulation Techniques. Proc. 3rd Indian Conf. on Computer Applications in Mineral Industry (ICCAMI-2001), New Delhi, pp301-314.

Sheorey, P.R. 1997. Empirical Failure Criteria, Publ. A.A. Balkema Rotterdem, p 176.

Straczek, John A. and Srikantan, B. 1960. The Geology of Zawar Zinc-Lead Area, Rajasthan, India; Memoirs of the Geological Survey of India, Vol 92, pp 12-46.

4

Numerical Stress Field Modelling for Underground Structures

Heinz Konietzky

ABSTRACT

Numerical stress field modelling in conjunction with stress field measurements has shown, that virgin in-situ stress fields are quite complex. The usual assumption of vertical principal stress is often not valid, especially in areas with complex topography and geology as well as regions, where tectonic forces act. Two examples are given, which demonstrate, that only the combined use of numerical 3D modelling and in-situ stress measurements can give a reliable picture about the stress state in the rock mass. These examples also show, that both, magnitudes and orientations of principal stresses can vary in a quite complex pattern even within short distances. The determination of the stress state is a pre-requisite for any subsequent reliable design and dimension of underground structures.

4.1 INTRODUCTION

Within the last decades numerical modelling has been developed as a standard tool in geotechnical engineering because of the following reasons:

- tremendous developments regarding the computer hardware, so that even very complex topics can be handled with time- and cost-efficiency on standard PC's.

- software developments, including pre- and post-processing, have lead to user-friendly codes.
- extensive validation and verification of the codes as well as numerous successful applications in the geotechnical praxis have lead to a broad acceptance of modelling results.
- the use of numerical models allows more cost effective engineering and analyses that are necessary to meet the standard of care requirements.
- numerical modelling gives insight into the real physics, which is invaluable for engineers involved in construction decisions.
- the applications of numerical methods is strictly recommended by latest standards in geotechnics (e.g. EUROCODE and Recommendations of the German Society of Geotechnis).

The advantage of numerical methods compared with analytical or empirical approaches is the physical handling of complex situations in respect to geometry, initial and boundary conditions, non-linear material, and hydro-thermo-mechanical coupled behaviour. Therefore, the primary stress state has a decisive influence on the overall model behaviour. The stress field modelling has to fulfil the following tasks:

- Determination of the principal stress directions.
- Determination of stresses outside of the area of measurements.
- Separation of local and regional stress components.
- Separation between topographic-gravitational, tectonic, and residual stress components.
- Proof of correctness of assumptions used in interpretation of stress measurements.
- Reduction of scatter obtain by stress measurements.
- Determination of stress changes due to secondary impacts (e.g. construction work).
- Understanding the history of existing stress fields.

4.2   APPLICATION IN RADIOACTIVE WASTE PROJECTS

4.2.1   Introduction

Describing the primary stress field in the region of a potential repository is an important aspect of a site characterization program. The primary stress field is a key input parameter for studies on the design and dimensioning of underground structures of any kind, as well as for calculating hydromechanically coupled processes and estimating long-term safety. Since, by nature, stress measurements always provide isolated results, model calculations are required. In order to maintain its natural integrity, direct measurements can only be performed outside the potential repository zone so that calibrated numerical stress field modelling becomes an essential element of site characterization.

4.2.2 Conceptual Model

The model has to meet the following conditions:

(i) Because of the complex tectonic-geological situation, a complete 3D model has to be selected.

(ii) The model must be capable of accurately reproducing the complex geological structure with several layers and fold structures.

(iii) The elasto-plastic behaviour of the rock layers and discontinuities must be reproduced using equivalent material laws.

(iv) The morphology of the Earth's surface has to be included in the model.

(v) The model must be sufficiently extensive in spatial terms to rule out artificial boundary effects and, at the same time, to take into account the influence of the nearby mountain range.

(vi) The model must take into account the additional influence of tectonic stress components.

(vii) Being part of an integrated site characterization program, the inner part of the model has to fit with the hydrogeological model in terms of model discretization and geological structure.

4.2.3 Numerical Model

The three-dimensional program 3DEC (Itasca, 1995) based on distinct elements (Hart 1995, Konietzky et al., 1994) was used for the numerical modelling.

The model domain was divided into three areas—a very finely meshed inner model area, a more coarsely meshed outer area, and a very coarsely meshed surrounding area. The total area covered by the whole model was around 300 km^3. The whole model consisted of 362.673 elements and its base was set at 3,000 m below sea level. The appropriate size of the numerical model was determined in advance by an extensive 2D model (Konietzky, 1995). The orientation of the outer model boundaries was based on the principal stress directions (main tectonic direction NW-SE). Fig. 4.1 shows the model structure with individual model areas marked separately, together with the locations of the boreholes in which the hydrofrac stress measurements used to calibrate the model were performed.

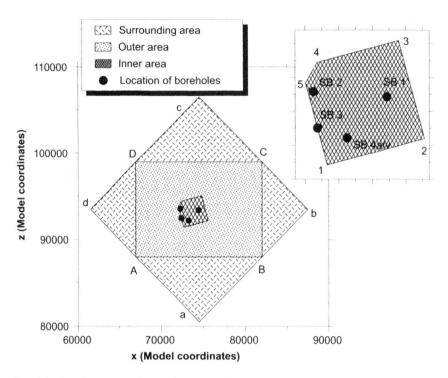

Fig. 4.1 Model structure of the individual model areas and locations of the boreholes where the stress measurements used to calibrate the model were performed (coordinates in m).

The discretization of the inner model area was done using the inner discretization scheme in a column structure (cuboid), which is similar to the existing hydraulic model. The geological structure of the inner model area, which includes the repository, is made up of a total of six different geological formations. Fig. 4.2 shows a 3D view of the inner model area. Fig. 4.3 also shows a 3D view of the inner model area, but with the covering layer removed, in a horizontal plane at the altitude of the potential repository (540 m a.s.l. [above sea level]).

Based on detailed laboratory investigations, a new material law was developed, tested, and implemented into 3DEC. The observed material behaviour is characterized as follows:

(i) The material has a non-linear failure surface.

(ii) The material has a marked strength anisotropy due to the existence of weaker bedding planes.

(iii) The material shows no notable strain-softening in the post failure region, but the initial strength values drop more or less abruptly to the residual strength values.

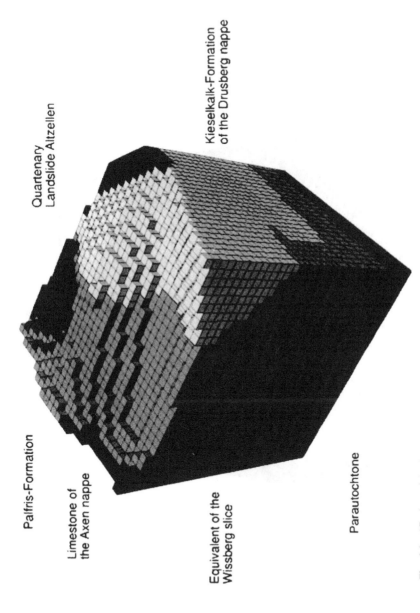

Quaternary
Landslide Altzellen

Kieselkalk-Formation
of the Drusberg nappe

Palfris-Formation

Limestone of
the Axen nappe

Equivalent of the
Wissberg slice

Parautochtone

Fig. 4.2 3D view of the inner area

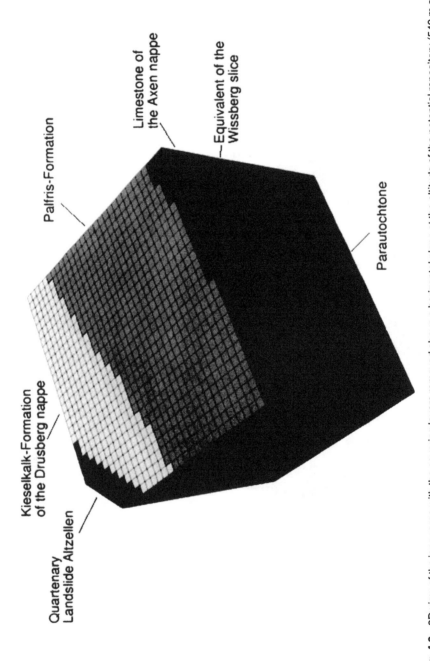

Fig. 4.3 3D view of the inner area with the covering layers removed above a horizontal plane at the altitude of the potential repository (540 m a.s.l.)

Based on the above mentioned characteristics, a new material law was developed, which is specified by the following features (see Fig. 4.4):

(i) The behavior of the matrix can be described by two bilinear Mohr-Coulomb yield surfaces with tension cut-offs. When the conditions on the virgin yield surface are met, it is replaced by the residual yield surface.

(ii) A ubiquitous discontinuity system is used, which can be described by two bilinear Mohr-Coulomb yield surfaces with tension cut-offs. When the conditions on the virgin yield surface are met, it is replaced by the residual yield surface.

A detailed description of the parameters used is given by Konietzky and te Kamp (1996).

The material model developed was generalized to allow the modelling of strain-softening, although this was not used in the presented work. The model needs specifications of at least 15 material parameters, which can easily be obtained from standard geomechanical testing. For the case of more detailed strain-softening behaviour, the variation of friction,

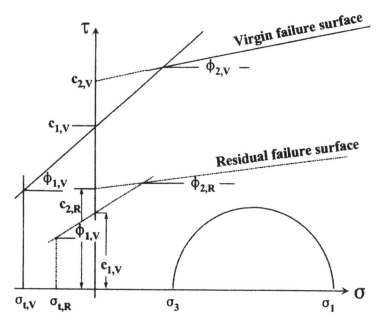

Fig. 4.4 Bilinear Mohr-Coulomb material law with tension cut off used in the 3D modelling. (τ: shear stress, σ: normal stress, c: cohesion, ϕ: friction angle)

dilation, cohesion, and tensile strength can be given as a function of plastic strain. Introducing additional tectonic compressive stresses, which increase with the depth and act in a NW-SE direction, led to a satisfactory agreement between measured and calculated results.

The results presented in detail in this report (variant T3) are based on the following approach for the horizontally acting, NW-SE-directed tectonic stress component: $S = So + dS/dz*Z$, where: z: depth, So: stress, dS/dz: stress gradient. For the range z = 0 to 500 m a.s.l. So = 5.000 MPa and dS/dz = – 10.000 MPa/km. For the range z = – 3.500 to 0 m a.s.l., So = 5.000 MPa and dS/dz = – 20.000 MPa/km.

Figure 4.5 compares the shut-in pressures from the hydrofrac stress measurements (minor principal stress) with the calculated stress-depth profiles along the tested boreholes SB1, SB2, SB3, and SB4a/v, assuming additional tectonic stresses. The location of the boreholes in the inner

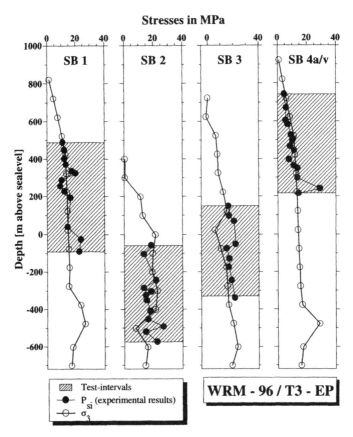

Fig. 4.5 Comparison of the shut-in pressures from the hydrofrac stress measurements with the calculated stress-depth profiles.

model area is shown in Figs. 4.1 and 4.9. Fig. 4.6 shows a detailed comparison of measured and calculated results for borehole SB3.

Material behaviour has a marked influence on the stress field. The following picture emerges for the level of the planned repository. Relatively stiff formations (Kieselkalk and Drusberg nappe and limestone

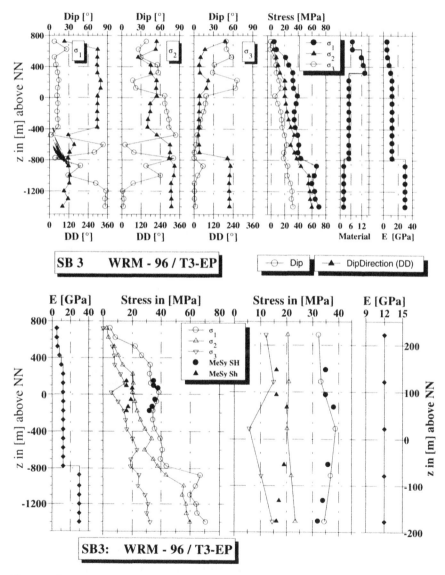

Fig. 4.6 Detailed comparison of measured data (S$_H$ and S$_h$) and calculated results for borehole SB3

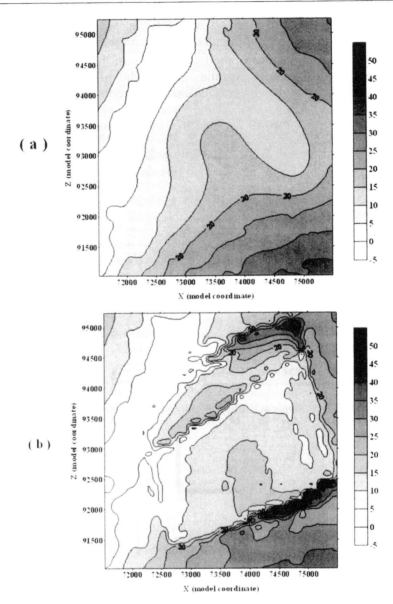

Fig. 4.7 Calculated vertical stresses at the repository plane for homogeneous parameters (a) and for realistic (inhomogeneous) geomechanical parameters (b).

of the Axen nappe) surround the soft marls of the Palfris formation. This leads to stress concentrations in the stiffer layers, which results partly in rock strength being exceeded (elasto-plastic stress redistributions). Figure 4.7 shows the results of calculations for the vertical stress component (assuming purely gravitational loading) for a horizontal section through the inner model at 540 m a.s.l. (potential repository level). Figure 4.7a shows the result of using a homogeneous elastic parameter set. In this case, the vertical stress reflects the features of the topography. Figure 4.7b shows the result with the same initial and boundary conditions, but using the real geomechanical parameters. The comparison clearly shows the influence different stiffnesses in the geological strata can have on the stress field.

An example of results for the modelling case T3 (elasto-plastic calculations) is given in Fig. 4.8, where the magnitudes and strike directions of the major principal stress for the inner model area are shown. The general trend of a NW-SE tectonic compression is clearly recognizable, but there are locally significant deviations. If the area is approximately 500 m a.s.l., the major principal stresses are mainly quasi-horizontal, whereas below 500 m they are vertical. Close to the surface the dip is determined mainly by the topography. Figure 4.9 shows the ratio of major to minor principal stress in the horizontal section 540 m above sea level with values of approximately 1.6 to 2.8.

Summarizing, the results of the numerical modelling and the main features of the stress field in the area of the planned repository can be described as follows:

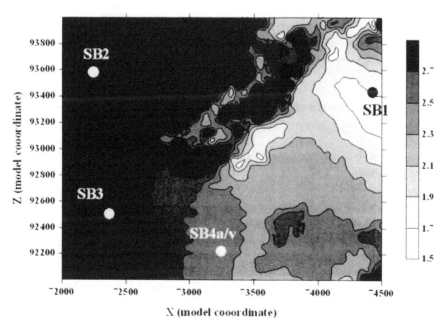

Fig. 4.8 Magnitudes and strike direction of the major principal stress for the inner model area

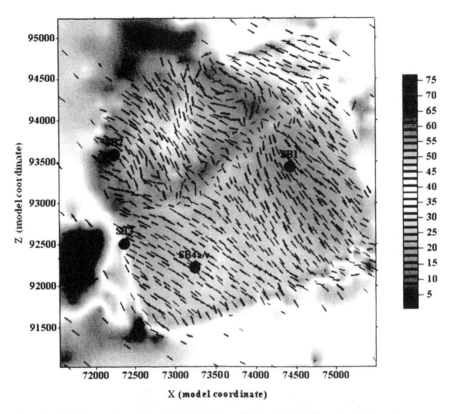

Fig. 4.9 Ratio of major to minor principal stress in the horizontal section 540 m above sea level (stresses in MPa)

(i) The in-situ stress field cannot be explained by the effects of gravity and topography alone. Modelling without tectonic stress components leads to magnitudes for the minor principal stress which were too low to explain the experimental data. Also, the directions were not consistent.

Considering the magnitudes and the orientation of the principal stresses, satisfactory results were obtained by applying a tectonic far-field component in the direction of N 135°. Three different tectonic models were tested and all three led to satisfactory results. The tectonic stresses applied to the outer boundary of the model, which led to the closest agreement with the observed stress field, varied bilinearly from 0 MPa at + 500 m to 5 MPa at sea level and 65 MPa at − 3,000 m. The results can be improved slightly by changing the tectonic far-field stresses, but the necessity for this seems to be questionable.

(ii) Above sea level (0 m), the major principal stress is approximately horizontal and strikes in the direction of about N 135°. The strike and dip of the intermediate and minor principal stresses vary.

(iii) At the level of the planned repository (540 m a.s.l.), the results are consistent with observed data. Strong variations in the resulting stress field coincide with material boundaries and regions where plastifications have taken place. Also, the ratio of the major to the minor principal stresses shows significant variations in plastified areas and at material boundaries.

(iv) In the area of the planned repository, the stress field is influenced by topography effects and does not show significant variations. The major principal stress is approximately horizontal and strikes in a direction of N 135°. The magnitudes of the principal stresses are within the following limits:

σ_1: 25 – 30 MPa σ_2: 14 – 18 MPa σ_3: 11 – 14 MP

(v) The ratio of the major to the minor principal stress is about 2.1 to 2.3 within the area of the planned repository site, where plastifications have not been observed in all modelling cases.

4.3 APPLICATION IN TUNNELLING

4.3.1 Introduction

This application describes three-dimensional numerical modelling studies of in-situ stress distributions in complex geological conditions. The modelling was intended to augment and generalize extensive hydraulic fracturing stress measurements carried out to aid in selecting the optimum alignment of an approximately 14 km long tunnel, part of a proposed new rail link between Stuttgart and Augsburg, Germany.

The numerical model includes specific representation of seven different geological layers and six geological faults with throws of up to 30 m. Results indicate complex and variable three-dimensional in-situ stress conditions along the tunnel routes. This is confirmed by the field measurements. Stress conditions are characterized by strong inhomogeneity and anisotropy with a maximum to minimum principal stress ratio of up to 4:1. The numerical model indicates a large change in orientation of the quasi-horizontal maximum principal stress direction along the tunnel route. This is also observed in the measurement results.

Based on the stress profiles from the model, the tunnel routes can be subdivided into four and five sections in each of which the stress conditions are approximately uniform. An initial assessment has been made of the necessary support measures and problems that may be anticipated during tunnel construction by determining a factor of safety for a circular tunnel of a certain diameter in each of the sections defined above.

Safe and economic construction and maintenance of tunnels depends on a detailed knowledge of geo-technical conditions along the proposed route of the tunnels. Planning and dimensioning of the tunnel has to consider both the hydromechanical characteristics of the rock mass and the loads that may be imposed on the supports. Detailed laboratory and field investigations of the mechanical properties of the rock mass are now considered to be a standard part of tunnel projects, but investigation of the in-situ stress field, on which the support loading depends, is rarely done. Stress measurements are considered to be expensive and difficult to interpret. Given the lack of reliable field data, support design is usually based on rough, over-simplified estimates of the stresses.

Use of in-situ stress measurements in conjunction with numerical modelling provides valuable advantages.

Thus,

- Stress measurements alone can provide data for individual points in space only. A numerical model is able to extrapolate individual values to larger areas.

- Stress measurements always determine a total stress. Numerical modelling is able to distinguish between the different components of the total stress, such as topographic, tectonic, and residual stresses. Also, numerical modelling allows the influence of geological variables (e.g. bedding planes, faults) on the stress field to be examined.

- The interpretation of stress measurement data is usually based on assumptions such as the relationship between deformation and stress. The influence of these assumptions on the estimated stresses can be examined by numerical stress field modelling.

4.3.2 Geology

The German Federal programme to upgrade and expand the national rail system ("Schienenwegeausbaugesetz") includes extension of the railway link Stuttgart – Augsburg in Southern Germany. This extension requires that the line ascend to the top of the Swabian Alb ("Albaufstieg"), a terrace with a difference in elevation of up to 400 m. To accomplish this, the line must be constructed within a tunnel.

Different routes for traversing this section have been investigated and rated with respect to geotechnical, hydrogeological, and ecological issues. Two alternative tunnel alignments were suggested; the so-called "optimized supplement alignment" ("optimierte Antragstrasse") and the

"Hasen valley alignment" ("Albaufstieg Hasental"). In both cases, tunnels of at least 13 km in length would be constructed. The "optimierte Antragstrasse" intersects the Alb scarp underground and continues in a tunnel to the Fils Valley. The Fils Valley is crossed via a bridge, approximately 90 m high and 470 m long, between Mühlhausen and Wiesensteig. The route then continues in a tunnel, reaching the top of the Swabian Alb southwest of Hohenstadt.

The alignment "Albaufstieg Hasental" is also a tunnel to the top of the Swabian Albs. The Hasen Valley, a side valley of the Fils Valley, is crossed via a bridge, approximately 40 m high and 160 m long, southwest of Wiesensteig.

The rock beneath the top of the Swabian Alb, consists of geological formations of the Brown (Middle) Jurassic and White (Upper) Jurassic period. Rocks of the Black (Lower) Jurassic period form the base. Quaternary sediments are also present, especially in the valleys. The Brown Jurassic formation consists of an approximately 270 m thick claystone sequence. Layers of sandstone occur intermittently, mainly in the lower and middle strata, whereas beds of limestone and iron oolites are found in the upper strata.

The White Jurassic formation is about 400 m thick and is composed of a banked interstratified limestone/marlstone in the lower and middle section and of partly banked, but mostly massive algal sponge and coral limestone layers. The massive limestone is strongly karstic in certain areas.

The Swabian Alb is part of the South German Great Block ("Süddeutsche Großscholle") which consisted originally of sediments up to 1600 m thick. During the transition from the Jurassic to Cretaceous periods, the region became dry; forming a mainland and was consequently subjected to erosion. Due to tectonic processes associated with the Alpine orogeny during the Cretaceous and Tertiary periods, the stratigraphic sequence became tilted. Thus, in the area of the ascent to the Swabian Alb, the Jurassic stratigraphic layers dip gently SE to SSE.

Tectonic deformation caused the formations to break along old, existing structural contours, and faults of various offsets were formed. As a result, a large number of small disturbances and fractured zones, striking mainly NE to SW with throws of up to 30 m, are encountered.

From the tunnelling perspective, an important consideration in rating the two alternate routes is, that high in-situ stresses may exist in the Brown Jurassic geological formation. The tunnel through these formations will be at a depth of up to 300 m. The stresses result, in part, from the depositional history of the general area and in part on local effects of geomorphology and tectonics.

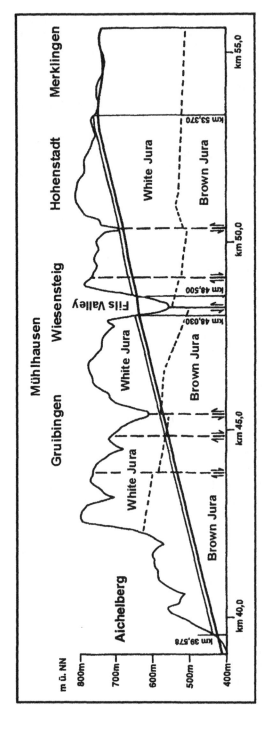

Fig. 4.10   Simplified geological longitudinal profile along the route "Optimierte Antragstrasse"

Figure 4.10 shows, schematically, a simplified geological longitudinal profile along the "Optimierte Antragstrasse" tunnel route.

Hydraulic fracturing tests to determine in-situ stresses in the region were carried out during the exploratory phase of the project. These results served as a basis for stress field modelling and further tunnel planning. The hydraulic fracturing stress results have been discussed in detail by Hammer et al. (1995). Figure 4.11 shows the location of the boreholes where the stress measurements were performed.

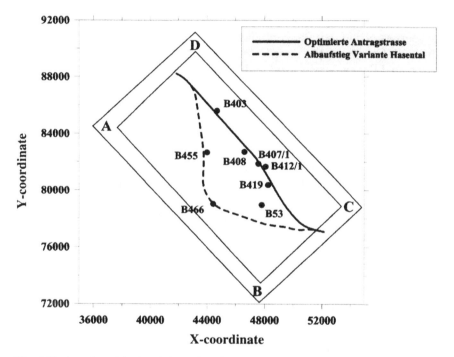

Fig. 4.11 Location of the boreholes, where hydraulic fracturing stress measurements were performed (including the numerical model boundaries and the tunnel routes)

4.3.3 Numerical Model

The three-dimensional distinct element code, 3DEC (Itasca, 1998) was used for the numerical modelling. This code is based on the Distinct Element Method (Hart 1993, Itasca 1998) and uses a time-marching calculation scheme. Calculation of stresses and deformations is based on the Explicit Finite Difference Method (see Konietzky et al. 2001).

The geological layers are simulated within a model with boundaries representing a region 17 km by 10 km in lateral extent and 1 km thickness, with a base at 200 m below sea level (– 200 m a.s.l.). The model consists of

2.559 deformable blocks and 662.606 elements (zones). The gridpoint distances within the inner model area, i.e. where the tunnel routes are located, vary between 20 and 70 m laterally and approximately 10 and 20 m vertically. The model includes a pronounced topography with differences in elevation of up to approximately 600 m.

The gravitational constant was set to 9.81 m/s^2. Normal velocities and displacements were prescribed on the outer vertical boundaries of the model. The base of the model was fixed in the vertical direction.

Since this stress field modelling exercise was carried out as part of the pre-planning phase of the project, data available on the mechanical properties of the various geological layers and the fault zones was very limited. Therefore, the simple Mohr-Coulomb material law with non-associated flow rule was used to describe the behaviour of the rock matrix and the Coulomb friction law for the fault zones. The dilation angle was set to zero, and the tensile strength was set to 1/10 of the cohesion. For the faults, the normal and shear stiffness were set to 1 GPa/m, the cohesion and tensile strength were set to zero and the friction angle was set to 20°.

The model contains six geological faults, dipping more or less vertically and striking E-W to NE-SW. Seven different geological layers were modelled (Figs. 4.10 and 4.12). At the faults, all of the known throws of up to 30 m were incorporated into the model. Exemplary, Fig. 4.12 shows the valley structures, where different geological layers outcrop.

The influence of the various model parameters was studied by conducting three very different situations:

   (i) a homogenous model under gravitational loading only;

   (ii) an inhomogeneous model including the geological layering and the faults under gravitational loading only; and

   (iii) an inhomogeneous model under gravitational loading with tectonic and residual stresses.

The model runs with gravitational loading only resulted in horizontal stresses much smaller than those indicated by the hydraulic fracturing stress measurements. It was necessary, therefore, to include the additional lateral (i.e. tectonic and residual) stress components. In the first such run, tectonic stresses only were added to fit the stress measurements. This was done by applying horizontal stresses, increasing linearly with depth (different gradients were considered), on the outer lateral boundaries of the model. Reasonable agreement with the measured stresses could be achieved, only when the model tectonic stresses were applied almost perpendicular to the direction of the actual tectonic stresses.

Since reasonable geological or tectonic explanation could not be found to support such a local rotation of the tectonic loading, the possibility of

Fig. 4.12 Detailed view of part of the whole model. Different shades indicate different geological formations.

this boundary loading was rejected. A more plausible explanation was needed.

Geological investigations had indicated that the initial rock cover in the region had reached approximately 900 m a.s.l. A lithostatic load equivalent to this height of overburden was therefore applied to the model. With subsequent erosion (reduction of rock cover and valley formation), the vertical stress decreased and uplift of the rock mass occurred. The model indicated that this decrease resulted in only minor decrease in the horizontal stresses. This was due to the fact that the outer boundaries of the model were fixed in the horizontal direction. This simulation was in good agreement with the stress measurements.

4.3.4 Results

This section discusses the results of the model finally accepted as most representative of the actual geological situation in the region.

Figures 4.13 to 4.15 show the stress profiles along the tunnel route "Optimierte Antragstrasse" (similar results were obtained for the alternative route, but will be not presented here due to the lack of space). The upper part of these figures shows the tunnel route together with the surface topography and the location of the hydraulic fracturing stress

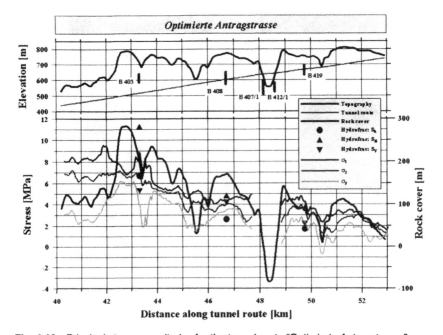

Fig. 4.13 Principal stress magnitudes for the tunnel route "Optimierte Antragstrasse"

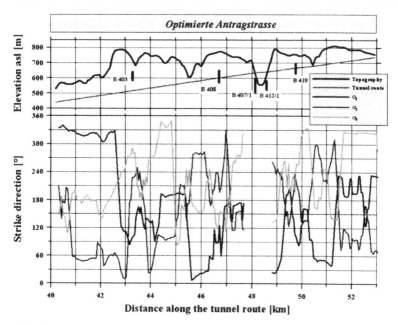

Fig. 4.14 Strike direction of the principal stresses along the tunnel route "Optimierte Antragstrasse"

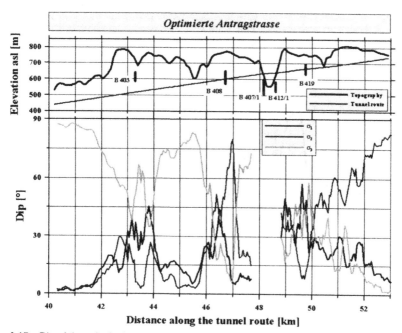

Fig. 4.15 Dip of the principal stresses along the tunnel route "Optimierte Antragstrasse"

measurement intervals. The lower part of the figures shows the magnitude, strike direction, and dip, respectively, of the principal stresses as calculated numerically, together with the measured values.

It is seen that, in general, the maximum principal stress is nearly horizontal. Except for the last 3 km of the tunnel route, the vertical stress component corresponds to the minimum principal stress. When approaching the valleys, the horizontal stress components acting perpendicular to the valley axis decrease, as expected. Due to the fact that the faults are nearly vertical and strike NW-SE, they have minor local influence only. The topography has a large influence on the stress field. In areas where the overburden is relatively high, the influence of the topography decreases and the ratio between maximum and minimum principal stress reaches values about 1.2:1. Where the tunnel approaches the surface or is near to the valley slopes, the stress ratio can go up to approximately 4:1.

Figure 4.16 shows the calculated ratio between the maximum and minimum principal stresses. As can be seen, the ratio varies strongly along the tunnel routes and reaches values varying between approximately 1:1 and 4:1. Both the measurements and the numerical calculations reveal a large rotation of the maximum quasi-horizontal stress component.

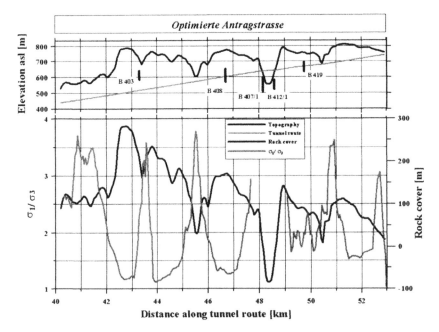

Fig. 4.16 Stress ratio between the maximum and minimum principal stress along the tunnel route "Albaufstieg Hasental"

Because most of the stress measurements were conducted along the "Optimierte Antragstrasse" tunnel route, comparison between measured and calculated values was possible for that route only. Figure 4.17 shows this comparison. Whereas the strike direction of the quasi-horizontal stress is nearly 0° (N-S) at 44 km it is seen to rotate to approximately 60° (NEE-SWW) at 53 km.

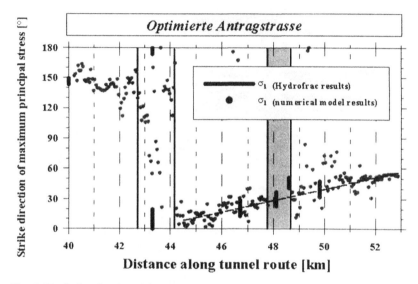

Fig. 4.17 Strike direction of the quasi-horizontal maximum principal stress along the tunnel route "Optimierte Antragstrasse"

In summary, the in-situ stress field to be expected along the tunnel routes will be non-uniform, varying considerably with principal stress ratios up to 4:1. The numerical model reproduces the measured values well and allows a reliable prediction of the stress field along the entire tunnel routes. The numerical model results show that the complicated in-situ stress field results mainly from the combined effect of the topography, the residual horizontal stresses due to the initial rock cover and the different stiffnesses of the rock layers. The fault systems have a secondary influence only. Fig. 4.18 shows the orientation of the quasi-horizontal maximum principal stress component for the two tunnel routes.

4.3.5 Practical Conclusions

During this preliminary stage of investigation (pre-planning phase), the main aim of the numerical stress field modelling is to obtain an initial idea of the rock stresses to be expected along the planned tunnel axis. These results can then be used for further planning of tunnel excavation methods

Fig. 4.18 Strike direction of the quasi-horizontal maximum principal stresses (short black lines) along the two tunnel routes investigated, together with the hydraulic fracturing measurement results (white dots with lines)

and the corresponding support measures. In the case considered here, the rock stress conditions along the tunnel routes can be used together with other criteria to decide on which of the two tunnel routes is preferable.

Based on the stress field modelling results, the two tunnels may be subdivided into 5 and 4 sections, respectively. Average values or gradients for the stress magnitudes and orientations can be given for each section, and used for further tunnel planning.

An initial assessment of where geotechnical problems may be expected along the tunnel route is provided by calculation of factors of safety according to the Mohr-Coulomb law for the various parts of the tunnels. Fig. 4.19 shows the factor of safety calculated along the tunnel route for several different combinations of cohesion and friction angle. Friction

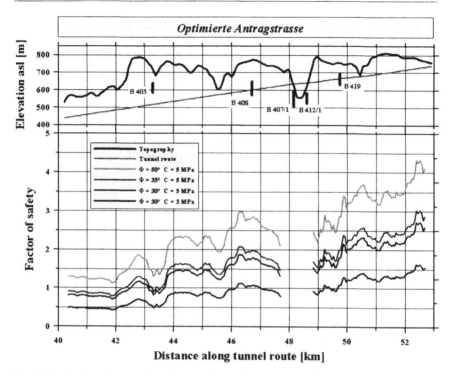

Fig. 4.19 Factor of safety along the tunnel route "Optimierte Antragstrasse" for several combinations of strength parameters

angles of approximately 30° to 35° and cohesion values between approximately 3 MPa and 5 MPa are expected along the tunnel route.

The numerical modelling has revealed the assumption that the vertical stress is a principal stress, is not valid at certain locations within the area under investigation. The determination of in-situ stresses from hydraulic fracturing tests usually makes this assumption leading, in this case, to incorrect values. It is recommended, therefore, that hydraulic fracturing measurements be re-analysed wherever the assumption of a vertical principal stress is likely to be violated.

Assessment of the numerical model results should recognize that, although a huge mesh with over 600,000 three-dimensional elements was used, the gridpoint distance is still on the order of several tens of metres. This means that the resolution in terms of stresses is on the order of approximately 0.1 MPa to 0.5 MPa only. Therefore, the accuracy and reliability of the very near surface stress values (i.e. at sections 50 km and greater) is limited. Also, the effect of the very local karst structures is not included in the model, nor is local material loosening and more complex material behaviour.

Comparison of the two alternative tunnel routes with respect to tunnel construction and considering the effect of the in-situ stress field only, allows the following conclusions to be drawn.

Both tunnel routes show significant anisotropy and inhomogeneity in the in-situ stress field, with principal stress ratios up to 4:1. Overall, the "Optimierte Antragstrasse" route shows slightly lower principal stress ratios and slightly higher safety factors.

Based on the stress field alone, the "Optimierte Antragstrasse" route is slightly preferable, although other criteria will probably play a much more important role in the final decision between the two alignments.

REFERENCES

Cundall, P. Formulation of a Three-Dimensional Distinct Element Model – Part I: A Scheme to Detect and Represent Contacts in a System Composed of Many Polyhedral Blocks, Int. J. Rock Mech., Min. Sci. & Geomech. Abstr., 25 (1988) 107, 116.

Hammer, H. Niedermeyer, S. Niedermeyer, Th. Untersuchungen zu Gebirgsspannungen und –bewegungen in der Schwäbischen Alb, Felsbau 6 (1995) 367-373.

Hart, R. 1993. An Introduction to Distinct Element Modelling for Rock Engineering, in: J. Hudson, ed., Comprehensive Rock Engineering, Vol. 2 (Pergamon Press, London) 245-261.

Hart, R.D. 1995. An introduction to Distinct Element Modeling for rock engineering. in: HUDSON: Comprehensive Rock Engineering, Vol. II, 245-261.

Itasca. 1995. 3DEC (Vers. 1.6), Minneapolis, Minnesota, USA.

Itasca. 1998 3DEC,Vers. 2.0, ITASCA Consulting Group, Inc., Minneapolis, Minnesota, USA.

Konietzky, H., Hart, R.D., and Billaux, D. 1994. Mathematische Modellierung von geklüftetem Fels. Felsbau 12, 395-400.

Konietzky, H. 1995. Three-dimensional stress field modeling of the Wellenberg site – Phase I. NAGRA Internal Report, NAGRA, p 42.

Konietzky, H., Blümling, P., and Rummel, F. 1995. In-situ stress field in the Wellenberg area. NAGRA Bulletin 26, 38-47.

Konietzky, H. Blümling, P., and F. Rummel. 1995. In-situ stress field in the Wellenberg Area, NAGRA Bulletin No. 26, 38-48.

Konietzky, H., te Kamp, L., Hammer, H., and Niedermeyer, S. 2001. Numerical modelling of in situ stress conditions as an aid in route selection for rail tunnels in complex geological formations in South Germany", Computers and Geotechnics, 28, pp 495-516.

Konietzky, H. and te Kamp, L. 1996. Three-dimensional stress field modeling of the Wellenberg site – Phase II. NAGRA Internal Report, NAGRA, p51.

te Kamp, L. Konietzky, H. and Blümling, P. Three-dimensional modeling of the planned Wellenberg repository site in Switzerland, in: Proc. NUMOG VII (A.A. Balkema, Rotterdam, 1999) 385, 390.

5

Application of Tunnel Boring Machines for High-Speed Tunnelling in Extreme Geological Conditions

McFeat-Smith

ABSTRACT

This chapter focuses on the application of tunnel boring machines (TBMs) for hard rock, mixed face, and soft ground conditions with potentially high water inflows, often within a single tunnel drive. The types of TBMs discussed are open-type hard rock with advance probing and grouting; the new range of sophisticated double shielded TBM; and dual mode EBPMs.

Field performance data in extreme mixed face conditions including tunnels in mountainous areas with high in-situ stresses and for sub-aqueous tunnel crossings are presented in terms of advance rates, utilization rates, and relative costs achieved on a range of projects throughout Asia. The prediction, site management, and effect on productivity of machines encountering major water inflows is highlighted.

Issues of maintaining acceptable levels of settlement and avoiding collapses to street level in urban area tunnels are also discussed together with the advantages and limitations of using earth pressure in comparison with slurry systems. A new, empirical method of selecting face control systems that can also be used for risk analysis is presented.

5.1 INTRODUCTION

The recent introduction and widespread use of TBM tunnelling into Asian Markets, most evident in Singapore, Hong Kong, the Philippines, and China has given rise to a dramatic fall in the cost of bored tunnelling. This has led to the undertaking of a series of highly ambitious projects where TBMs are employed to excavate through soft ground, extreme mixed face conditions, and hard rock often with potentially high water inflows in mountainous, sub-aqueous and intensely developed urban areas.

Results being achieved from such installations vary from being highly successful to disastrous, sometimes on adjacent sites, highlighting the need for improved site investigation, planning, and better methods of selecting machines for the specific ground conditions encountered.

Experience and data is drawn in this paper from a very wide range of projects, geological conditions, and TBM types as identified in Table 5.1. In order to maintain confidentiality these projects are not necessarily referred to in the text.

5.2 TBMS FOR HARD ROCK TUNNELLING

5.2.1 Selection of TBMs Based on Advance Rates and Costs

A series of case histories of successful applications of open type and double shielded TBMs types in hard rock of UCS strengths up to 460 MPa have been presented by McFeat-Smith (1998); and McFeat-Smith, Grandori, and Concillia, (1999).

Table 5.1 *TBM Project List*

Project	Location	Geology	TBM (diameter)
Strategic Sewage Disposal Scheme	Hong Kong	Sub-aqueous & urban; volcanic tuffs, granites	Open type 3 – 5 m
Tai Po to Butterfly Valley Aqueducts	Hong Kong	Mountainous tuffs, granitic	Open type 3 – 5 m
Tolo Effluent Export Scheme	Hong Kong	Mountainous granitic	Double shielded 3 – 5 m
Silvermine Bay Aqueduct	Hong Kong	Mountainous tuffs, granitic	Double shielded 3 – 5 m
KCRC DB320 West Rail	Hong Kong	Urban granitic, mixed, soft	Dual mode 8.75 m

(Table 5.1 Contd.)

(Table 5.1 Contd.)

Project	Location	Geology	TBM (diameter)
Kai Tak Sewage Transfer Scheme	Hong Kong	Urban granite, mixed, soft	Dual mode slurry 5.6 m
MRT NEL Contract 704	Singapore	Urban granite, mixed, soft	EPBM 6.5 m
MRT NEL Contract 705	Singapore	Urban old alluvium, marine	EPBM 6.5 m
MRT NEL Contract 710	Singapore	Urban soft, mixed, sedimentary	EPBM 6.5 m
Powergrid Cable Tunnel	Singapore	Sub-aqueous, karst soft, mixed, sedimentary	Dual mode 6.0 m
Deep Tunnel Sewage Scheme T-01 to 4	Singapore	Urban old alluvium	EPBM 6-7 m
DTSS T-5 to 6	Singapore	Urban soft, mixed, granites	EPBM 4-6 m
Casecnan Power Tunnel	Philippines	Mountainous sedimentary, volcanic	Double shield 5.0 m
Umiray-Angat Power Tunnels	Philippines	Mountainous sedimentary, volcanic	Double shield 5.0 m
Guangzhou Metro, Line 2	China	Urban soft, mixed, sedimentary	EPBM 6.5 m
New Wu Chieh Power	Taiwan	Mountainous slates	Open type 6.0 m
Melen Aqueduct Bosphorous Crossing	Turkey	Subaqueous & urban sedimentary, karst	Shielded 6.0 m

From this database and other projects, Fig. 5.1 has been drawn up to highlight typical advance rates achieved against the full range of IMS ground classes (McFeat-Smith, 1998) for open type, double shielded, and dual mode hard rock TBMs.

In simplified terms the conditions represented by IMS ground classes are listed below:

IMS Ground Class	Geological Condition
1	Massive, few joints
2	Favourable rock
3	Moderately fractured or weathered
4	Highly fractured or weathered
5	Fault zones or completely weathered rock
6	Soil conditions

The pattern of peak performance of open-type TBMs in IMS classes 1 and 2 is logical, this falling off is rapid as the ground condition deteriorates and heavier support measures are required. Double shielded TBMs, in comparison, achieve their peak performance in more fractured rock masses (class 3) and give better performance in poorer rock mass conditions due to the use of the TBM shield and segment erecting facility.

The pattern of advance rates against ground conditions for the dual mode hard rock TBM is similar to that of the double shielded TBM. Advance rates are subdued by access constraints, particularly for interventions for cutter changes but are very high in favourable soil conditions where no mixed face or boulder conditions are experienced.

The pattern of relative costs/m of tunnel for excavation and support costs for different IMS classes shown below (for the double shielded TBM only) vary unsympathetically when plotted against the advance rates, this after residual set-up costs have been accounted for. This pattern is similar for most tunnelling methods.

From the data provided in Fig. 5.1 it is evident that the selection of TBM type for a particular rock tunnel can be based upon establishing the expected ground conditions (rock hardness to assess instantaneous penetration rates and IMS ground class to predict utilization) and predicting TBM advance rates and costs.

A notably advanced approach to TBM selection was adopted by GLF/ SELI for the 13.2 km long, 4.88 m diameter water tunnel for the Umiray Angat Irrigation and Power Scheme in the Philippines. The TBM was driven from a single portal using a purpose designed, double shielded Robbins TBM, and included the following features:

- high cutterhead power (1890 kw)
- high torque (2,367 knm)
- variable frequency drive
- shield diameters progressively reduced from front to rear to cope with converging ground
- a telescopic shield design allowing opening up of the shield to access rock close to the tunnel face
- a back-up system designed to be compatible with the TBMs capabilities
- use of honeycomb precast segments to allow simultaneous excavating and lining installation.

As described by McFeat-Smith (2000) and Grandori (2001) the tunnel was driven through a series of poorly investigated, water bearing, volcanic/sedimentary terrain with over 1000 m cover, high in-situ stresses and major fault zones.

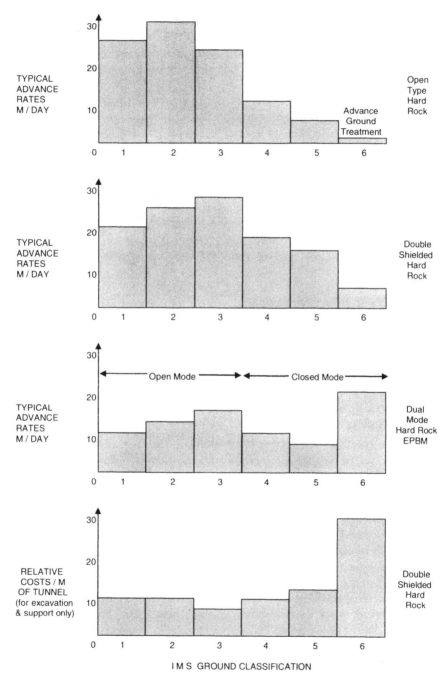

Fig. 5.1 Typical advance rates and relative costs for 5 – 6 m diameter open type, double shielded, and dual mode TBMs

For the first 5 km the TBM was serviced by helicopter for transportation of parts and construction materials.

Several major fault/collapse zones were encountered requiring special support measures facilitated by a layout allowing direct access through the TBM cutterhead.

Severe convergence also occurred up to a maximum of 0.37 m in competent rock, requiring excavation by hand through the telescopic shield to free the cutterhead in several locations. High inflows of up to 200 lt/sec with cumulative inflows of over 850 lt/sec had to be managed on site by means of diverting inflows.

The hard rock tunnel was excavated in 24 months at an average of advance (including all events) of 528 m in a month. This was achieved by:

- use of an advanced, specifically designed double shielded TBM and back up system, and
- using a well organized and motivated management and labour force.

5.3 WATER INFLOW ISSUES

5.3.1 Prediction

Prediction of water inflows into rock tunnels depends upon the local topography and geology, and from works by McFeat-Smith, McKean, and Waldmo (1998) depends upon:

- the size of the water source (including catchment zones within adjacent hillsides).
- the head of water above the tunnel.
- the horizontal separation between the water source and tunnel.
- the recharge to the water source.
- the degree of joint openness.

Patterns of water inflow in individual tunnels have been found to be relatively consistent against IMS rock classes as illustrated on Fig. 5.2. The inflow data in each class have been averaged over some 5 km of tunnel. The general trend is logical, such that water inflows increase with the degree of fracturing in rock masses and, in the case of IMS rock class five fault zones may be filled with relatively impermeable gouge materials or in weathered rock zones with clayey materials. Hence, from a combination of assessed IMS classes, permeability testing, and the hydro-geological factors mentioned above it is often possible for an experienced engineering geologist to build up a reasonably accurate estimate of water inflows into rock tunnels. Clients should nevertheless appreciate that this is an imprecise art and that anomalies may occur.

5.3.2   Management of Water Inflows

Whenever possible, rock tunnels in potentially wet conditions (particularly sub-aqueous) should be up-gradient drives, as the cost of transfer using pipelines, pumping, emergency back-up facilities, and maintenance of rolling stock can be considerable for down-gradient drives.

On the Hong Kong SSDS project extensive advance probing and grouting was undertaken in extremely strong granites and tuffs at 10-11 bar water pressure to meet strict water-tightness specifications (Grandori, Concillia, and Nardone, 2001). For example, an increase in total inflow rate from about 3 to 10 lt/min/m resulted in more than 70% reduction in TBM utilization as illustrated in Fig. 5.3, which again has been averaged from data averaged over some 5 km of tunnel. Overall, a utilization of 33% was achieved in non-sensitive tunnel sections on Contract DC/96/17 compared with 14% in sensitive urban areas. Nevertheless high settlements (up to 800 mm) were recorded on site.

Experience gained also demonstrated that post-excavation grouting was of little value until the tunnel lining could be installed in that no containment system existed to prevent water inflows transferring to other parts of the tunnel. Conclusions to be drawn from this are:

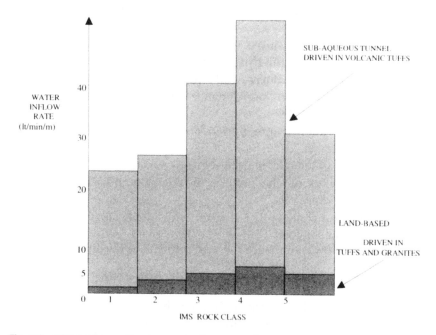

Fig. 5.2   Water inflow pattern in massive igneous rocks

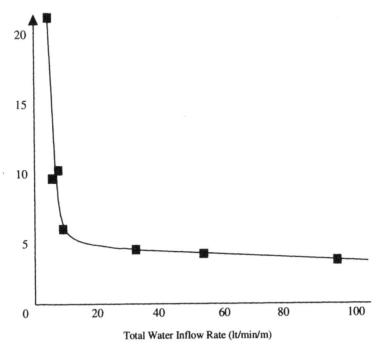

Fig. 5.3   Correlation between total water inflow rate and TBM utilization

- the use of EPB or slurry machines would not have been practical for this project as full hydrostatic pressures were realized on site.
- advance grouting ahead of advancing hard rock TBM drives has a major impact on performance and achieves uncertain or detrimental results.
- post-excavation grouting prior to lining installation is likely to be ineffective.
- the high settlements monitored were related as much to the total water inflows into the tunnel as the thickness and sensitivity of the overlying soils.
- the early installation of a watertight lining, albeit from a non-watertight shielded TBM, would have greatly reduced the overall water inflows and hence the high settlements encountered.

5.4   TBM FACE PRESSURE CONTROL SYSTEMS

While many pipejacking slurry applications have been undertaken, with major applications of slurry machines in Japan and for the Sydney Airport Link, experience to date in SE Asia has been mainly with EPB systems.

This being largely due to concerns about cost and the lack of space for soil separation facilities.

5.4.1 Mixed Face Conditions

EPBM applications in extreme mixed face conditions have taken place in Hong Kong and Singapore in extremely strong granitic type rocks overlying completely weathered granite and other soils as illustrated previously in Table 5.1.

Table 5.2 provides a rationalization of this experience in a range of IMS ground classes and mixed granite face conditions for a 7.5 m diameter hard rock EPBM operating in a closed mode using compressed air or even grouting to facilitate disk cutter interventions. The very high downtime necessary to achieve cutter changes at 2-3 bar compressed air pressures are evident as are the lower rates of advance in mixed face conditions.

Table 5.2 *Typical tunnel advance rates for 7.5 m diameter EPBM operating in closed mode*

TBM Activity	IMS Ground Classes					Mixed
	1-2	*3*	*4*	*5*	*6*	*Faces*
	Competent granite		*(Fractured)*	*(Faults)*	*(Favourable /Soils)*	*>30% Rock*
Delays for Segments %	6	8	10	12	14	8
Cutter Interventions %	45.9	43.1	43.3	34.8	32	40.6
Downtime for Soil Disposal %	8	9	10	11	12	13
TBM % Maintenance/Probing	8	9	10	11	12	13
Survey/Water/Ventilation %	3	4	5	5	6	7
TBM Utilization %	29.1	26.9	19.7	26.2	24.0	18.4
Instantaneous Penetration Rate (m/hr)	1.5	2.0	2.3	1.9	4.0	2.0
Advance Rate (m/week – 120 hr/week)	52.4	64.6	54.4	59.7	115.2	44.3

High rates of advance are however achieved in soils without boulders. In sedimentary terrain higher rates of advance is achieved using EPBMs with similar power and torque, and would also be expected using slurry systems due to reduced delays for cutter interventions.

5.4.2 Ground movement and control issues

Settlement is an inevitable consequence of tunnelling works and could be in the order of 100 - 120 mm in close vicinity of deep cut and cover tunnels. For the Hong Kong MTR compressed air tunnels ground settlement measured directly above the tunnel line in the order of 50 mm was not uncommon.

Settlements recorded above EPB tunnels driven in Singapore and Hong Kong are listed below.

Table 5.3   *Maximum ground settlements recorded directly above 6.5 –9 m dia. EPB tunnels*

Strata	Settlement in mm	
	Range	Average
Old Alluvium	– 6 to 18	8
Mixed Marine/Old Alluvium	25 to 67	44
S2 & S4 Sedimentary	10 to 28	20
Mixed faces S2/S4	14 to 42	30
Completely Weathered Granite (CWG)	15 to 25	17
Granitic Rock Types	0 to 2	1
Mixed CWG/granite	10 to 33*	20*

*Also see comments below with respect to collapses

In comparison, movement on building structures have been very low, often in the order of 1 to 2 mm.

Figure 5.4 shows maximum settlements monitored above an EPBM drive in Singapore's Old Alluvium together with corresponding face pressures. In this case, as the EPBM progressed to Ring No. 131 mixed faces of this partially cemented sandy silt and overlying soft marine clays were encountered and the face pressure was lost. This gave rise to a steady increase in settlement from 11 to 25 mm and eventually to 67 mm. The contractor stepped up the face pressure from 50 kPa as shown to 350 kPa before the ground movement was eventually arrested and reduced to normal.

This near incident demonstrates the risks of operating EPBMs at low pressures in the vicinity of such mixed face conditions. Factors found to enhance the maximum settlements in EPBM drives include:

- mixed face conditions.
- old stream courses.
- thicknesses and properties of soil troughs.
- slow progress, e.g. during learning periods and to negotiate site constraints such as breakthroughs into station boxes.

In more extreme mixed face conditions the risk of loss of the EPB plug becomes very high, demanding constant vigilance by experienced operatives. Such conditions may be created by large competent boulders as well as partial faces of competent hard rock and soils. Also water inflows can be high at such interfaces. Della Valle (2001) describes problems experienced in Singapore, which can lead to excessive cutter and cutterhead wear, progressive collapse of the upper soil strata, and

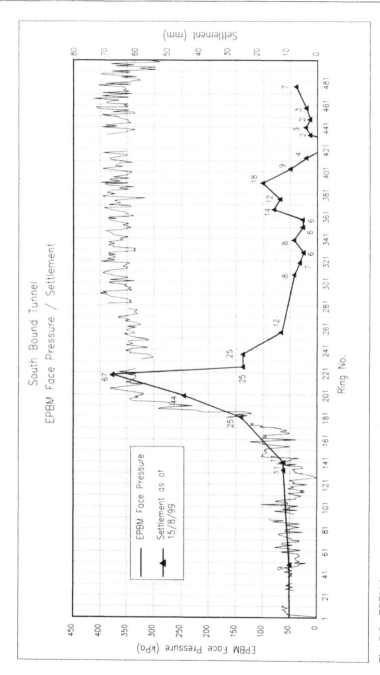

Fig. 5.4 EPBM face pressure versus ground settlement

chimneying of the voids up to street level. In such extreme conditions the extensive use of bentonite and foaming agents is not always sufficient to arrest collapses, which can be in the order of 30-100 m³ at a frequency of about 1/200 m in such mixed face tunnelling.

The alternatives appear to be the use of ground treatment, preferably in advance of tunnelling from the surface and/or the use of slurry systems despite the higher overall cost and difficulties of separation. The latter is highly desirable for subaqueous drives and cutter changes can be both lower and more easily facilitated as less time is needed to clear the cutterhead chamber of spoil.

5.4.3   Selection of Face Control System

Figure 5.5 shows the IMS method of selecting face control systems for soft ground TBMs from particle size and SPT N values. This has been drawn up by the writer from a series of case histories such as that shown in Fig. 5.4, from discussions and data from leading contractors and manufacturers and hence is based upon collective experience from the industry. It is presented as a general guide rather than an alternative to knowledge of the local geological conditions. Users are also advised to consider the local hydro-geological conditions as existing or potential new aquifer zones may require a higher level of control.

Clearly, the amount of fines and degree of soil stiffness are considered key factors for selection and use of open and low EPB pressure and grey areas exist where both EPB and slurry systems can be employed. For extreme mixed face conditions, such as those described previously, the writer recommends that slurry is considered first and alternatives selected only if adequate site investigations (or preventative measures) are available to justify the use of EPB systems.

5.5   CONCLUSIONS

This chapter presents performance data and information from a series of TBM installations that are fundamental for understanding how ground conditions affect, and are affected by different types of modern TBM types in Asia, often operating in extreme geological conditions.

Empirical patterns of TBM advance rates, costs, water inflows, and face control requirements for the full range of IMS ground classes have been developed and are being applied for tender and risk assessment purposes. This approach is essential for selection of appropriate TBM types and ground control systems given the very risky nature of the installations now being undertaken throughout Asia.

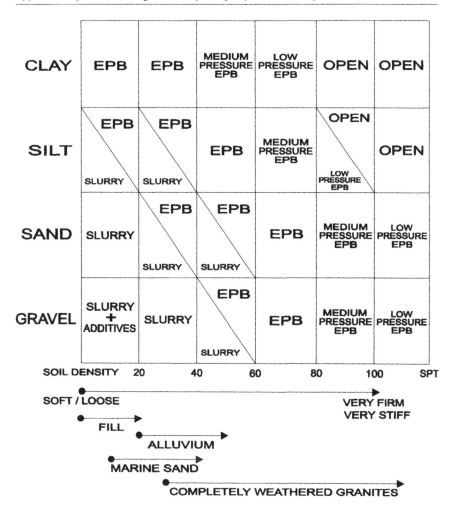

Fig. 5.5 IMS Method of selecting face control for soft ground TBMs

The advantages and limitations of the TBM types and systems available for tunnelling in such extreme conditions as mixed faces of competent granites and soft soils have been discussed and the case history from the Umiray Angat tunnel in the Philippines clearly demonstrates the benefits that can be gained by specialist TBM contractors using specifically designed, modern TBMs for such projects.

It is hoped that with continuing feedback data and specific methodology of the type illustrated in this chapter, coupled with improved tender planning, TBM options, and their control systems can be selected to achieve less risky, more improved performance on site.

REFERENCES

Della Valle, N. 2001. Boring through a rock-soil interface in Singapore. RETC Proceedings, San Diego. SME Inc. 633-645.

Grandori R. 2001. Manila Aqueduct (Philippines) – The Construction of the Umiray-Angat Tunnel Project. RETC Proceedings, San Diego. SME Inc. 777-790.

Grandori R., Concillia, M., and Nardone, P. 2001. Hong Kong Strategic Sewage Disposal Scheme (SSDS) TBM Tunnelling Beneath Seabed and Urban Areas. RETC, San Diego. SME Inc. 743-754.

McFeat-Smith, I. 1998. Mechanized Tunnelling in Asia. Published IMS Tunnel Consultancy Ltd. Hong Kong 1998 www.imstunnel.com.

McFeat-Smith, I., MacKean, R., and Waldmo, O. 1998. Water inflows in bored rock tunnels in Hong Kong: Prediction, construction issues and control measures. ICE Conference on Urban Ground Engineering, Hong Kong.

McFeat-Smith, I., Grandori, R., and Concillia, M. 1999. Construction of Hong Kong Governments First Two Land-Based TBM Projects, Proc. of 3rd Asian Tunnelling Summit, IBC. Hong Kong.

McFeat-Smith, I. 2000. Breakthrough in the Philippines, Part 1 June, Part 2 July Issues, Tunnels, and Tunnelling.

6

Bhuj, 2001 and New Madrid, 1811 – 1812: The Case for Analog Earthquakes

Michael Alexander Ellis

ABSTRACT

Almost every aspect of the 1811-1812 New Madrid earthquakes appears to fly in the face of seismological conventional wisdom. The inferred moment magnitude from ground motions (including liquefaction) is significantly larger than permitted using reasonable stress drops and rupture areas inferred from current seismicity. Surface deformation is not accumulating in patterns or magnitudes that are consistent with paleoseismological evidence of recurrent large earthquakes. All of this has inspired a level of debate that is, or was prior to the 2001 Bhuj earthquake, stalemated by a lack of quality observations. The magnitude (M) 7.7 Bhuj earthquake, in the state of Gujarat, India, now provides the opportunity to calibrate and understand many of these conflicting or unusual observations. Ultimately, the Bhuj earthquake should yield a significant improvement to seismic hazard models of both the continental USA and subcontinent of India. This chapter provides a summary of the important characteristics of each of these and related earthquakes. The most significant findings to date are that in all likelihood the Bhuj earthquake ruptured deep into the lower crust, involved a large stress drop, and was driven largely by compressional stresses related to the convergence of India and Eurasia. There is no compelling evidence that the Bhuj earthquake is representative or an extension of the diffuse India-Eurasian plate margin to the west. These characteristics provide credence to the suggestion that the New Madrid earthquakes ruptured significantly deeper than the nominal 15 km to which current seismicity extends and that in all likelihood, stress drops during these earthquakes were relatively large. Finally, I suggest that we drop the terms intraplate and interplate, because they carry with them fossilized implications that do not reflect the complexities behind the origin and significance of earthquakes.

6.1 INTRODUCTION

On the Indian Republic Day, January 26, 2001, a moment-magnitude M 7.7 reverse-fault earthquake occurred near Bhuj in the state of Gujarat. Official Indian government figures recorded a human death toll of a little under 20,000 and economic loss at about \$1.3 billion (Bendick et al., 2001).

The Bhuj earthquake occurred about 400 to 600 km away from the nearest plate boundary[1]. Strong-ground motion was felt over a region from Madras to Kathmandu and Calcutta, distances of over 2,000 km and more than 16 times that of the magnitude 7.8, 1906 San Francisco earthquake. These observations alone have prompted the suggestion that the Bhuj earthquake is analogous to the infamous New Madrid earthquakes, a series of three great earthquakes that occurred over a three-month period in the winter of 1811-1812 (Ellis et al., 2001). If such an analogy is appropriate, the Bhuj earthquake offers an unparalleled and for all practical purposes, a unique opportunity to better quantify seismic hazard in both the continental USA and the subcontinent of India.

In this chapter, I explore the proposition that the New Madrid and Bhuj earthquakes are analogs. Three characteristics of an earthquake address the issue of analog behaviour in the context of seismic hazard. The first is the nature of the rupture itself, which primarily involves scaling issues and stress drop. (A related issue but one that is beyond the scope of this chapter is the associated aftershock decay pattern, which carries important consequences for seismic hazard.) The second involves the nature of the crust through which seismic waves are propagated and

---

[1]The range given here is because of the shallow fold-thrust belts that spill onto the Indian subcontinent along all of the Pakistan part of the plate boundary. These are rooted at some unknown distance farther west, closer to the Chaman fault, and it is here that the important and relevant boundary conditions exist.

attenuated. Third is the tectonic setting of the earthquake: how was it driven and the corollary, when will it happen again?

6.2 THE BHUJ AND NEW MADRID EARTHQUAKE RUPTURE CHARACTERISTICS

The Bhuj earthquake: State of knowledge, early 2002

The Bhuj earthquake occurred on January 26, 2001, at approtimately 23.5°N, 70.2°E. Source parameters, derived from the Harvard CMT catalog and the NEIC are listed in Table 6.1. The most notable feature is the apparent lack of surface rupture, despite the relatively shallow depth (Wesnousky et al., 2001). Compare this to the smaller 1992 magnitude 7.3 Landers earthquake, which generated approximately 85 km of surface rupture (Sieh et al., 1993), or to the 1999 magnitude 7.6 Chi Chi earthquake that ruptured almost 100 km of the surface. Standard scaling relations for intraplate earthquakes (Wells and Coppersmith, 1994) suggest that we should have seen almost 200 km of surface rupture from the Bhuj event.

The distribution of aftershocks (recorded by imported local seismic arrays) shows clustering of events at depths from 6 to 35 km and along what are interpreted as the lateral tip zones of the rupture, defining and enveloping a roughly trapezoidal plane that dips and narrows to the south at about 40°(Raphael et al., 2001; Negishi et al., 2001). (Note that this plane is considerably steeper than either of the relevant nodal planes derived from teleseismic data, see Table 6.1)

Table 6.1 *Source parameters for the Bhuj earthquake from the Harvard CMT catalog (hrvd) and the USGS National Earthquake Information Center (NEIC). M_0 is $\times$ 10^{20} N_m.*

Time (UTC)	M	M_0	lat	long	depth	nodal planes	source	P-axis (trend)
3:16:54	7.6	3.4	23.63	70.24	19.8	298/39, 66/64	hrvd	177
3:16:40	7.7	3.8	23.419	70.232	20.0	283/33, 115/74	NEIC	175

The size of the inferred rupture plane is about 40 km by 40 km, relatively small for an earthquake of this size. Inversions of teleseismic data for slip suggest a relatively simple asperity near the hypocenter with a maximum slip of 10 m (Mori et al., 2001) to 12 m (Antolik and Dreger, 2001). Antolik and Dreger (2001) also suggested a second subevent at shallow depths with slip up to 6 m. Thick basin sediments above the hypocenter, however, produced reverberations in the P-waveform, which Langston (2001) suggests may be misconstrued as contributions from shallow faulting.

An estimate of average slip may be derived from the definition of seismic moment, $M_o = \mu Au$, where M_o is seismic moment, μ is the average shear modulus taken to be 3.3×10^{10} Pa, A is rupture area, and u is average slip. Using the range of M_o in Table 6.1, the average slip lies between about 6.4 and 7.2 m, consistent with estimates of maximum slip from inversion of teleseismic data (Mori et al., 2001; Antolik and Dreger, 2001). An estimate of the static stress drop, $\Delta\sigma$, may be derived from static crack models (Eshelby, 1957) which takes the general form for a circular crack of $\Delta\sigma = C\mu(u/r)$, where C is a constant equal to 7/16, and r is the crack (or rupture) radius, here approximately 20 km. Assuming u = 7-10 m, the static stress drop is about 16 to 23 MPa. Even the low end of this estimate is anomalously high, and the high end is more than twice typical values (Hanks and Johnston, 1992). The value in parentheses (u/r) is half the shear strain drop associated with the rupture, which here yields a value of about 1.75×10^{-4}, an order of magnitude higher than typical values for thrust earthquakes (Scholz, 1990). These estimates of scaling and stress drop are important in understanding the patterns of associated strong ground motions and likely form a common tie to the 1811-1812, New Madrid earthquakes in the central USA.

The 1811-1812 New Madrid earthquakes: State of knowledge, 2002

It is obviously difficult to offer a direct comparison to source parameters of the New Madrid earthquakes, because these are derived from a historical and generally less understood data source. Nonetheless, some New Madrid source parameters are well-founded, usually at the upper limit. A range of moment-magnitude estimates for the largest of the three earthquakes, which is most likely to be the third and only thrust earthquake in the sequence, is 7.5-8.1, derived from the Modified Mercalli Intensity scale, which maps areas of ground shaking to magnitude (Johnston, 1996; Hough, 2000). Three features of New Madrid seismicity offer bounds on the surface area of each rupture: map pattern, tip-zone earthquakes, and the depth distribution of earthquakes. The map pattern of seismicity offers a relatively well defined left-stepping strike-slip fault system. If we assume that the present seismicity tracks the underlying active faults, then the maximum length of any individual rupture is likely to be less than about 140 km. Two relatively large historical earthquakes (M 5.5 to 7.2) have occurred at the southwestern and northeastern ends of the seismic zone, consistent with the presence, at these extremes, of a tip zone that is propagating to southwest and northeast, respectively. Present seismicity is limited to the basement underlying Paleozoic and younger sedimentary rocks but shallower than about 15 km (Chiu et al., 1992).

We have no constraints on the average slip for either of the first two earthquakes, because neither is known to have surface rupture. The third earthquake ruptured the surface within the step-over region and probably where the displacement was highest, but again constraints are poor (Johnston and Schweig, 1996). The scarp associated with this thrust event is compound and known to have existed in part prior to the 1811-1812 earthquakes (Kelson et al., 1992). Trenching studies show the scarp to be associated with two pre1811-1812 earthquakes. A best guess, therefore, for displacement during the third New Madrid earthquake is about 3 m.

These dimensional bounds infer a maximum seismic moment, assuming that displacement scales with the equivalent radius of the rupture, $r^* = (LW\pi)^{1/2}$ (where L and W are length and down-dip width of the rupture, respectively) of approximately 2.08×10^{20} Nm (M 7.5). This estimate is well below that derived from the area distribution of ground motion (Johnston, 1996). If we assume instead a strain drop similar to the Bhuj earthquake of about $1.75 \times 10^{-4} = u/r^*$, then $M_0 = (1.75 \times 10^{-4}), \mu LWr^*$ $= 3.14 \times 10^{20}$ Nm (M 7.6). An alternative solution, also suggested by the Bhuj earthquake and proposed earlier by Johnston (1996) is that the depth of rupture was significantly greater than present seismicity would indicate. In order to find the width of rupture that can generate a M 8.1 earthquake, which scales as the equivalent radius, the definition of seismic moment is solved for the width:

$$W = (1/L).[M_0 \sqrt{\pi} /k.\mu]^{(2/3)} \tag{1}$$

Where k is the scaling relation, $k = u/r^*$. Thus, a magnitude M 8.1 requires a width of about 48 km, and a M 8.0 earthquake requires about 38 km. Solving for k instead and asking, what scaling is implied by generating a M 8.0 if the rupture reaches depths of 33 km (a depth used by Johnston, 1966)? Rearranging [1] to solve for k gives:

$$k = (M_0 \sqrt{\pi} /\mu).[1/LW]^{(3/2)} \tag{2}$$

This yields a value of about 2.15×10^{-4} to generate a magnitude M 8.0 and an implied average displacement of only 2.6 m, which is remarkably consistent with the earlier estimate of displacement across the Reelfoot fault scarp.

The Bhuj and 1811-1812 New Madrid earthquakes appear to have in common a lack of surface rupture, a relatively small rupture area for the magnitude of the earthquake and by inference, a relatively large stress drop. This might be considered an overly incautious statement, because there are large uncertainties surrounding the New Madrid earthquake parameters. It seems inescapable, however, that the New Madrid earthquakes must have ruptured deeper than the nominal 15 km to which current seismicity extends, and that this happened during the Bhuj earthquake adds some weight to the likelihood during the New Madrid events. Adding further weight still is the observation that the equally large earthquakes of 1819 Rann of Kachchh (Bilham 1999) and 1897 Assam (Bilham and England, 2001) also failed to rupture the surface but may have ruptured at the base of the crust. I argue that the estimates of strain drop (the scaling factor, k) made above are reasonable first-order estimates. The extent to which such strain drops necessarily lead to high stress drops, however, depends on the scaling processes. These are largely unknown for earthquakes within plates and remains a high research priority in the current U.S. Geological Survey's National Earthquake and Hazards and Reduction Program.

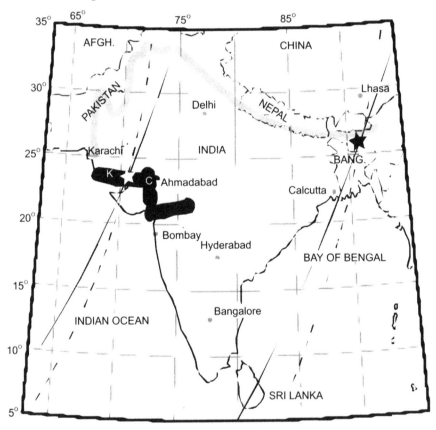

Fig. 6.1 Location of the Bhuj earthquake *(white star)*. The Narmada-Cambay rift complex is shown by the solid lines, letters identify the individual rifts: K, Kutch; N, Narmada; C, Cambay. The continuous and dashed lines that sweep across the map are small circles about Euler poles for the motion of India with respect to Eurasia derived from, respectively, the joint inversion of seismicity and Quaternary fault slip rates (Holt et al., 2000), and Nuvel1 (DeMets et al., 1990). These are shown passing through the Bhuj epicenter and through the epicenter for the Assam earthquake of 1897, which is shown by a black star north of Bangladesh. The Indian-side of the diffuse plate boundary between India and Eurasia is shown by the grey lines.

This sketch is not a political map and does not purport to depict political boundaries. The rough political delineations shown here may not be correct or accurate.

6.3 DISTRIBUTION OF GROUND MOTION: THE MMI ISOSEISMALS

The second ingredient to any seismic hazard analysis and the second important characteristic of an earthquake is the distribution of tremors. Close to the source, ground tremors are strong and may exceed 1 g in exceptional circumstances. Strong ground motion of a fluid-saturated granular material, particularly if capped by a relatively impervious and strong layer (such as a clay-rich soil), can lead to extensive ground failure through the process of liquefaction. Field investigation and analysis of satellite imagery has revealed up to 15,000 km^2 of liquefaction as a result of the Bhuj earthquake (Tuttle et al., 2001). In comparison, the 1811-1812 New Madrid earthquakes generated liquefaction over about 10,000 sq km. Exposures in the Bhuj region, however, show individual liquefaction features (feeder dikes in particular) to be up to a few tens of cm in width, in stark contrast to the meter-plus dimensions of New Madrid features. Liquefaction features are often the only paleoseismologic tool in the investigation of the New Madrid earthquake chronology, and one of the perennial challenges for paleoseismologists is to quantify the relation between the extent and style of liquefaction and large earthquakes. The Bhuj earthquake offers, yet again, a unique opportunity to address this relationship.

One of the immediate consequences of the Bhuj earthquake was the vast area over which it was felt. Fig. 6.2 shows at the same scale, the felt areas of both the Bhuj and the December 18, 1811 New Madrid earthquake. For comparitive purposes, Fig. 6.2b also shows MMI isoseismals from the 1992 M 7.3 Landers earthquake. The Landers earthquake was reported to have an anomalously high stress drop of about 20 MPa (Sieh et al., 1993), and yet the isoseismals are clearly significantly smaller than those during the December, 1811 New Madrid earthquake. The high-stress drop of Landers is considered to be related to the long-recurrence interval of events on this fault plane (Sieh et al., 1993), whereas the relatively small felt area is probably caused by the physical state of the crust. Seismic waves that propagate through the southwestern part of the USA need to travel through the southern Basin and Range province, a region active extension, relatively abrupt changes in crustal thickness, and high heat flow. These conditions differ markedly from both the Indian subcontinent and the bulk of the remaining North American continent, where the crust has remained relatively inactive for hundreds of millions of years, heat flow is in general low or normal, and the Moho has relatively little relief. It would seem that high-stress drop alone, therefore, is insufficient to generate the large felt areas typical of intraplate earthquakes.

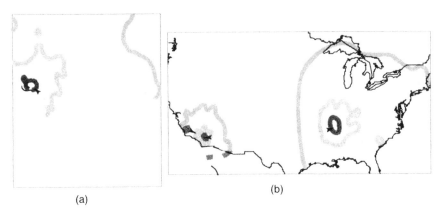

(a)

(b)

Fig. 6.2 Felt areas as quantified by the Modified Mercalli Intensity scale of the January 26, 2001 Bhuj, India earthquake (a), the December 18, 1811 New Madrid earthquake and the 1992 Landers earthquake (b). These contours shown are approximate only and were modified from Hough et al., (2000) and from the USGS facility, ShakeMaps (http://www.trinet.org/shake/)

If high-stress drop is related in part to the healing time allowed a rupture zone, then one test of the importance of stress drop to the generation of large felt areas within continents is to re-examine felt areas from earthquakes along the Himalayan front. Himalyan thrust faults accommodate at least 12 mm/y and up to about 20 mm/y via large earthquakes that probably recur every 500 years. Seismic waves from these events will travel through the same crust as did waves from the Bhuj and other earthquakes within the Indian subcontinent. In this context, Molnar's (1987) remarks on the distribution and intensity of ground motions during the 1897 Assam earthquake are interesting. He states, p.24, that "if steeply dipping planes were involved (in the Assam rupture), there must have been several of them for such a widespread distribution of high intensity."

The New Madrid and Bhuj earthquakes are clearly analogous in terms of ground motions. This analogy has critical consequences, because the quantification of seismic hazard also depends on the spectral content of the passing wave train. This is because different types of infrastructure are sensitive to different parts of the spectrum. Spectral content depends on both source (magnitude and stress drop) and attenuation characteristics of the crust, but these relations are poorly known for large earthquakes and short distances. The Bhuj earthquake has offered the opportunity to quantify these relations in conditions that are as close to the central USA as we are likely to find, at least in our lifetimes.

6.4   BHUJ AND NEW MADRID: THE CASE FOR A TECTONIC ANALOG

A tectonic analog requires that the driving forces for any particular suite of earthquakes or active deformation are essentially the same. At face value, the tectonic setting of the Bhuj and New Madrid earthquakes appears remarkably similar. They both occupy an ancient rift system, occur in a rigid plate, and suffered underplating by a hot spot. In this section, I briefly summarize the geologic setting and the current and paleoseismicity of each region, again emphasizing, as needs be, the similarities and contrasts. This will provide a context for evaluating the potential driving forces and perhaps resolve the proposition that these are tectonic analogs.

The Bhuj earthquake occurred near the southern boundary of the Kutch[1] basin, one of the three Mesozoic rifts that forms a complex system of rifts or grabens at the edge of the Indian continent (Biswas, 1987), including the ~NS trending Cambay rift and the EW trending Narmada rift. The Kutch basin reflects the earliest rifting of Gondwanaland and is filled with approximately 3 km of late Triassic to lower Cretaceous sedimentary rocks. Most of these were deposited on a Precambrian granitic basement, currently exposed as topographic highs in what is otherwise a monotonously flat landscape, during transgressive and regressive cycles in the Middle and late Jurassic, respectively. These deposits were later folded during the late Cretaceous and subsequently overrun by basalts of the Deccan Trap at the end of the Cretaceous. The Narmada basin appears to have formed later than the Kutch basin, but of the three it alone records marine conditions during the late Cretaceous. The entire region was eventually exposed to non-marine deposition in the late Cretaceous, probably in response to uplift associated with the Reunion plume. Reinterpretation of analog seismic refraction data gathered in the 1980s supports the existence of a large and relatively shallow mafic pillow under the rift complex (Kumar et al., 2000), which may be the relic of the Reunion plume head. Interestingly, the Cambay basin (also Mesozoic in origin, with up to 1200 m of sedimentary rock beneath the Deccan Trap basalts), subsided faster during the Tertiary as ~ 5 km of sediments were deposited (Biswas, 1982). The origin of this subsidence may be related to subsequent cooling and sinking of the mafic pillow.

By comparison, the general setting of the New Madrid earthquakes appears relatively simple. The New Madrid earthquakes and current seismicity occupy the Reelfoot rift, a single NE-trending graben about 250

---

[1]Kutch is the alternative (and apparently modern) usage for Kachchh. The terms are interchangeable, and the author uses both terms here in accord with the prevalent use for the particular feature.

km in length that presently occupies the upper Mississippi embayment in the central USA (Ervin and McGinnis, 1975). The Reelfoot rift initiated during the late Precambrian, during the formation of Iapetus. Reelfoot rift is generally interpreted as an aulocogen or failed rift, one arm of a triple-rift junction the remaining two of which evolved into the mid-Iapetus rift. The rift is well defined by both gravity and magnetic potential fields (Hildenbrand, 1985). Recent analysis of gravity data strongly supports the existence of a relatively dense mafic rift pillow sitting at lower crustal to shallow upper mantle depths (Stuart et al., 1997). Approximately 5 km of post-rift sedimentary rocks are deposited across the rift. During the Mesozoic, sediments were deposited on a southeast-sloping continental shelf. At about 70 Ma, the Bermuda hot spot is thought to have perturbed the region sufficiently to have generated a series of mafic intrusions that have punctured the rift structure. The definition of the rift structure has been emphasized by the formation of the Mississippi embayment, a broad downwarping that began in the Paleocene and ended sometime in middle Tertiary. The origin of the embayment is unknown.

Seismicity within each of the rift systems is relatively high and each has sustained a number of moderate to large historical earthquakes. The Kachchh region includes the well known Narmada seismic zone, which centres largely on the upper Cambay rift. The seismic zone is characterized by frequent small-magnitude earthquakes, and nine moderate (5<M<6) and damaging earthquakes have occurred within the last 170 years, including the M~6.1, 1956 Anjar earthquake (Bendick et al., 2001). Seismicity does not appear to define neat linear faults, but appears relatively scattered within the seismic zone. The largest historical earthquake is the M~8, 1819 Rann of Kachchh (also known as the Allah Bund earthquake), which generated a 6 m high 6 km long natural dam to the south-flowing Puran (also Nara) river (Baker, 1846; Oldham, 1898). Importantly, the Rann of Kachchh earthquake, like the Bhuj event, did not appear to rupture the surface. Bilham (1998) analysed Baker's (op. cit.) original levelling survey of the Allah Bund scarp and suggested that the earthquake slipped 11.5 ± 1.5 m on a ESE-striking, NNE-dipping reverse fault that terminated 300-600 m below the surface. The axis of principal shortening from both the large 1819 and 2001 earthquakes and virtually all moderate earthquakes within the Narmada seismic zone are roughly N to NNE. Although these earthquakes occur within an earlier rift complex, they do not accommodate rifting today. Active deformation is entirely compressional and consistent with both the relative plate motion between Eurasia and India (Fig. 6.1) and the modern convergence within the Indian subcontinent (Paul et al., 2001; Bilham and Gaur, 2000).

Microseismicity in the NM region, in contrast to that in the Narmada rift complex, appears to delineate a series of faults at depths ranging from

5 to 15 km. The most prominent alignments of seismicity, however, do not outline the rift bounding faults. Instead, seismicity outlines a relatively well defined left-stepping strike-slip fault system (Russ, 1979). The few focal mechanisms so far constrained indicate a predominantly strike-slip sense of displacement across all but the central zone of seismicity. The centrally located left-step is marked by a region of uplift of no more than about 10 m, which forms the hanging wall of the reverse Reelfoot fault. Focal mechanisms in this central uplifted zone are mixed, probably because they occupy a volume that is within the tip zones of several strike-slip faults as well as within the hanging wall of the reverse fault. Consequently, the strain field within the central zone is likely to be significantly heterogeneous. The New Madrid region has also suffered from a series of moderate (4<M<6) earthquakes in the last few decades. Two of these occurred at the respective ends of linear seismic zones, which may be evidence of propagating tip zones. Prior to the great 1811-1812 New Madrid earthquakes, there is now a voluminous amount of evidence from studies of liquefaction that severe strong ground shaking earthquakes, probably as large as the 1811-1812 sequence, occurred in about 1400 AD and in about 900 AD (Tuttle et al., 1999).

Faults of the New Madrid seismic zone, at least those that are suspected to have ruptured during the 1811-1812 sequence have been shown to accommodate a regional strain that is consistent with the regional in-situ stress field (Ellis et al., 1995). The regional stress field in turn is consistent with ridge-push forces derived from the mid-Atlantic rift boundary (Richardson, 1992). Importantly, as for active deformation in the Kachchh region, the Reelfoot rift while containing the bulk of seismicity in the central USA, is not actively rifting.

Despite the significant deformation within these regions, each also occupies a relatively rigid plate. The rigidity[1] of the Indian plate, determined from GPS measurements, is about 1.5 to 7×10^{-9} (Paul et al., 2001; Bilham and Gaur, 2000). The central USA is considered rigid at about the same level of compliance (T. Dixon, personal communication, 2001).

It is worth emphasizing the similarities so far described. Each of the Bhuj and New Madrid earthquakes sit within an ancient rift system, but modern deformation is contractional, not extensional, and shortening axis are consistent with ridge-push (in the case of New Madrid) and the relative motion of colliding plates (in the case of Bhuj). Each rift system appears to be associated with a large mafic pillow derived most likely from hot spots under but not associated with the older rift system. And each region sits within a relatively rigid plate.

---

[1]The author uses (and prefers) the definition of rigidity used in the geodetic community, which uses the straightforward and direct measure of strain rate.

Beyond these similarities, however, important differences in boundary conditions probably exist. The Rann of Kachchh and Bhuj earthquakes are about 400-600 km from the nearest plate boundary in southern Pakistan. It is clear that while structures and seismicity in both the Kachchh and plate boundary regions (Bernard et al., 2000) accommodate the relative motion of India and Eurasia, they do so very differently. Active deformation along the plate margin is strongly partitioned between largely strike-slip earthquakes within the interior of the diffuse orogeny and classic shallow fold-and-thrust structures that spill as festoons onto the Indian plate. By contrast, seismicity and structures within the Kachchh region appear to be deeply rooted, forming a relatively simple opposing pair of crustal-scale reverse faults. The Kachchh region may also be influenced by thrust loading to the northwest and north, which is in part responsible for the upper-Indus depocentre in central Pakistan. A further complication is that the Indian plate may be buckling, albeit evidence is ambiguous and the theoretical possibility seems remote (Bendick and Bilham, 1999). In contrast, the New Madrid region is a thousand kilometres and more from any plate boundary, the most obviously influential of which is the mid-Atlantic rift, whose ridge-push forces may be recognized in in-situ stress patterns across the USA (Zoback, 1992).

The delicate balance of forces to which these and any actively deforming region is subject will likely be perturbed by climate changes over time scales that are sufficient to significantly affect the restribution of mass. In the case of advancing and retreating glaciers, this time scale may be as short as a thousand years. This may be pertinent to the New Madrid region, which sits at the outer bulge of the last significant glacial advance and which has been offered as a means to initiate deformation in the region (Grollimund and Zoback, 2001; Pollitz, 2001), although Wu and Johnston (2000) suggest the affect is insignificant. The most likely result of climate change in the Bhuj region is the filling up of the Rann of Kachchh and probably large parts of the lower Indus basin. The effect of this will possibly be a temporary slowing of deformation, but over what time scale is beyond the scope of the current paper. The point of my raising this is to emphasize the complexities that very likely exist within and at the boundaries of these regions. This is a necessary caveat to carry the readers into the next section, in which I describe the present tectonic models for the Bhuj and New Madrid earthquakes.

Driving the Bhuj-Kachchh and New Madrid earthquakes

If the Bhuj and, by inference, the 1819 Rann of Kachchh earthquakes and the New Madrid earthquakes are tectonic analogs, the driving force for each should be comparable. There is no argument that the New Madrid earthquakes lie well within the North American plate, and that their origin must owe something to local conditions. But this much alone does not preclude the possibility that plate boundary forces provide the driving energy. By the same token, the proximity of the Bhuj and Rann of Kachchh earthquakes to their local plate boundaries does not preclude the possibility that driving forces are generated completely locally.

I begin with models proposed for the New Madrid region, because these are generally more fully developed and cover a range of possibilities that may be directly relevant to the Bhuj and Rann of Kachchh earthquakes.

Tectonics models of the NMSZ fall into two basic classes: the standard San Andreas plate boundary model (e.g., Weber et al., 1998; Newman et al., 1999) and non-plate boundary models (e.g., Grana and Richardson, 1996; Stuart et al., 1997; Gomberg and Ellis, 1994; Grolimund and Zoback, 2001; Kenner and Segal, 2000; Pollitz et al., 2001). The critical distinction is that non-plate boundary models do not require large displacements in the far field; elastic strains are generated locally, most likely by buoyancy forces derived from density contrasts in the lower crust or upper mantle (e.g., Grana and Richardson, 1996).

These recent models have been motivated explicitly by two important observations: 1) a wealth of paleoseismological evidence, summarized in Tuttle et al. (1999) that persuasively attest to a sequence of strong ground shaking events (most likely earthquakes) limited to the past ~10,000 years, and 2) the apparent absence of significant modern deformation (Newman et al., 1999). In response to these observations, Kenner and Segal (2000) proposed that a low-viscosity region in the lower crust of uncertain origin could be concentrating and delivering stress into the upper crust. Coseismic failure across upper crustal faults is then thought to re-energize the lower crustal body, which in turn delivers stress to the upper crust, and so the cycle continues until the system runs out of energy. This model explains well the finite time span of earthquake activity in the region, and these authors state that "large earthquakes can occur every 500 to 1000 years, even though surface deformation rates are less than the detection level of prior GPS surveys, approximately 1×10^{-7} year^{-1}. This means that existing geodetic data cannot be used to rule out the occurrence of future damaging earthquakes in the NMSZ." By prior GPS surveys and existing geodetic data, these authors are referring to the campaign style survey and data used by Newman et al., (1999).

The Kenner and Segal (op. cit.) model did not provide clear predictions of interseismic surface velocities, nor did it explain the origin of the posited lower crustal low-viscous body. Pollitz et al. (2001) have effectively extended this model by proposing that a mafic rift pillow, described and characterized by Hildenbrand (1985) and Stuart et al. (1997), began sinking in response to the last period of deglaciation. These authors provide a set of predictions for interseismic surface velocities that involve primarily radial subsidence (although there are potential complications with post-seismic effects). Pollitz et al. (op. cit.) suggest that the GPS data of Newman et al. (1999) go some way to corroborate such a radial subsidence.

Although there are inevitable devils in the details of these models, they illustrate very well the need for high-quality GPS data. If driving forces for earthquakes in the New Madrid region are generated locally, surface deformations will need to be carefully resolved in order to properly and responsibly evaluate the seismic hazard. The Center for Earthquake Research and Information at the University of Memphis has recently finished the installation of a permanent 12-station GPS array in mid-America (GAMA), which is centred over the New Madrid seismic zone. Preliminary results from GAMA, however, do not show a radial subsidence to support the theoretical predictions of Pollitz et al. (op. cit.).

Note also that the translation from surface velocities to slip rates and then to recurrence interval is critically dependent on the manner in which strain is accumulated (Kenner and Segal op. cit.).

In summary, it is probably fair to conclude that GPS data in the New Madrid area are not consistent with our current understanding of plate boundaries (Newman et al., 1999), which is perhaps not surprising. But this is probably as far as current GPS results will take us. We cannot yet distinguish between the various mechanisms proposed in recent years, and importantly, we cannot yet say that driving forces are generated locally or from a far-field plate boundary. This important distinction bears on the issue of recurrence and longevity of the New Madrid fault system as a contributor to the seismic hazard of the central USA (Schweig and Ellis, 1994).

There are far fewer tectonic models for the Bhuj and the Rann of Kachchh earthquakes, but the more complex tectonic setting will surely generate a similar number of different models as there are and have been for the New Madrid earthquakes. At issue here is whether the driving forces are analogous to those that drive the New Madrid earthquakes.

An elegant model has been proposed by Li et al., (2002), in which deviatoric stresses are concentrated in the Kachchh region through the sum of stresses derived from the segment of the plate boundary that runs through Pakistan. It is also assumed that the adjacent oceanic crust is about 10 MPa stronger than the continental crust and that the Kachchh rift is about 10 MPa weaker than the surrounding crust. Stresses arise from the Pakistan part of the boundary because it is assumed to be partially locked, in accord with the large moment-deficit along this part of the boundary (Bernard et al., 2000). Li et al., (op. cit.) show the orientation of principal stresses at a depth of 20 km to be consistent with the Bhuj and other regional mechanisms. However, the orientation of these stresses is largely dictated by the model plate compression, which was designed to yield a homogeneous shortening of 2×10^{-9}/yr, consistent with the GPS results from Paul et al., (2001). This input is essentially the propagation into the plate of far-field plate boundary forces, and it represents the net relative motion of India with respect to Eurasia. The influence of the simulated partially locked western boundary in Pakistan appears minimal in the results of Li et al. (op. cit.). The more critical components of the model appear to be the stronger oceanic plate and the assumed local rheological weakening, both reasonable processes to assume. The influence of the western plate boundary is perhaps the most difficult process to simulate, largely because the physical significance of the moment-deficit is unknown (is the deficit made up by aseismic processes, and is it representative of longer time scales?), and because, as Li et al., (op. cit.) pointed out, the nature of the western boundary is probably quite complex (e.g., Haq and Davis, 1997). Li et al. (op. cit.) conclude that the origin of the Bhuj earthquake and seismicity in the Kachchh region is due to the combination of plate boundary processes and local rheological weakening.

An alternative tectonic model suggests that the Kachchh region can be likened to a small platelet or rigid inclusion (called the Sind Block) that is in the process of being carved out of the Indian subcontinent and transferred into the diffuse plate boundary that lies between Eurasia and India (Stein et al., 2001). In this model, the Bhuj (and presumably the Rann of Kuchch) earthquake is on the southern boundary of the Sind Block (Stein et al., 2001). The eastern boundary of the Sind Block turns north and northwest and heads for the tip of the Sulaiman Lobe across the lower Indus basin. This model is not without its own analogs. Stein et al., (op. cit.) liked the process to the behavior of the Sierra Nevada block, which moves independently from the enveloping North American plate. There are also well known rigid units within the Eurasian diffuse boundary, including the much larger Tarim block and the smaller Katawaz block that lies between the Chaman fault and the Sulaiman ranges (Haq and Davis, 1997). Nonetheless, this model remains fairly conjectural. There is no obvious evidence of any deformation along the longer eastern boundary of the Sind Block, although it is possible that this might be hidden by alluvium of the Indus River.

With the present state of knowledge, I agree, in spirit if not in detail with Li et al., (in press) that the driving force for the Bhuj and regional earthquakes is the sum of plate boundary stresses coupled with a local rheological heterogeneity. I differ from their conclusions, however, that the western boundary conditions, at least as they have characterized them, are important. The clearest manifestation of the boundary conditions along the western boundary are the festooned fold-thrust belts themselves, and the extent of that direct influence is given by the dimension of these fold-thrust belts. Farther inland (or inplate), the origin of deviatoric stresses are probably predominantly derived from far-field compression from the relative motion of India and Eurasia. Other processes, mainly rheological contrasts, may cause variations in the delivery of that compression, but I suggest that the fundamental origin remains the far-field plate compression and that the westen boundary is largely irrelevant.

Driving forces, plate rigidity, and recurrence intervals

An understanding of the driving forces (the tectonic model) responsible for earthquakes is not simply an academic exercise; the connection to seismic hazard is important. Understanding the driving forces sets the methodology to better evaluate both the theoretical estimates of recurrence interval (via strain accumulation and recurrent relaxation) and the field observations of paleoseismicity. Getting the model wrong, or prematurely dismissing the right model (e.g., Newman et al., 1999), can demonstrably lead to equally wrong evaluations of seismic hazard (e.g., Newman et al., 2001).

If far-field plate compression drives earthquakes in the Bhuj region, then unless plates are infinitely rigid, which they are not, any level of compliance will ultimately provide sufficient deviatoric stress to induce brittle failure, so long as the rate of strain accumulation is faster than can be accommodated through aseismic processes. Given this last caveat, the level of plate rigidity (or the ambient rate of strain accumulation) then provides us with an upper bound on a long-term recurrence interval.

Plate rigidity is difficult to measure, because the relevant line-length changes are lower than processing errors of geodetic data. That is, errors in data processing provide only upper bounds to rigidity. Within the Indian subcontinent, for example, the residual associated with line-length changes between Delhi and Kanyakumari (a north-south distance of about 2000 km) between 1994 and 1999 is approximately 3 mm/yr, equivalent to a strain rate of about 1.5×10^{-9}/yr. Other estimates of the strain-rate, using different groupings of the geodetic stations in the processing, give values from $5\text{-}7 \times 10^{-9}$/yr.

In the case of the Bhuj earthquake, if we take the conservative estimate of plate rigidity (1.5×10^{-9}/yr over a line length of 2000 km) but place all displacement in the Bhuj region (3 mm/yr), then enough shortening will occur to yield a Bhuj or Kachchh-type earthquake (about 4.5 m to 5 m, respectively, of horizontal shortening) every ~1500 and 1700 years. If we use instead a value of rigidity of 5×10^{-9}/yr, the implied recurrence intervals are 450 and 500 years. Rajendran and Rajendran (2001) document from liquefaction data a pre-Kachchh earthquake between 880 AD and 1035 AD. If we assume from this result and for illustration a recurrence interval of 1000 years, then half of the compliance of the Indian subcontinent would be needed to generate earthquakes of Kachchh or Bhuj size. If the compliance is isotropic and of the order of 3 mm/yr, then there is sufficient energy to generate once in every 1000 years a Bhuj-sized earthquake anywhere in the Indian subcontinent, so long as the total contraction along a single azimuth (effectively a small circle of latitude about the India–Eurasia Euler pole) is no larger than 3 mm/yr. In this context, it's important to note that the Rann of Kachchh and the Bhuj earthquake occurred on adjacent piece of an Euler-latitude, and the 182 years that separates these earthquakes has little to do with the recurrence interval on either of the two fault segments. (I say "little" because there are complications induced by stress transfer that may physically couple adjacent fault segments.) I use the figures above largely for illustration, because very little is known about the extent to which strain is inhomogeneous within India, and data to evaluate rigidity is very sparse. It is clear, however, that dense geodetic control over the entire Indian subcontinent (and probably starting with the Bhuj-Kutch region) would be an essential and very useful means to help constrain estimates of long-term earthquake recurrence.

Returning to the central USA, there are far more geodetic data available both regionally and within the New Madrid area and estimates of rigidity and strain are in general better constrained. GPS processing residuals are strongly correlated with the data length (time span) and indicate that the plate rigidity is bounded by GPS processing errors of 1-2 mm/yr. These bounds would provide sufficient energy to drive large earthquakes every one to few thousand years. But in the New Madrid region, local velocity differences (over length scales of tens to at most a few hundred of kilometers and an approximately 7-year time span) are not permitted to be greater than about 3-5 mm/yr (Newman et al., 1999). In other words, where the wholescale plate rigidity permits sufficient displacements, it does not seem to be accumulating locally. The problem is compounded by independent and incontrovertible evidence that the recurrence interval of strong-ground shaking events is of the order of about 500 years (Schweig et al., 2001; Tuttle et al., 1999). The possibility of higher strain rates localized over the most active parts of the seismic zone exists given the measurement uncertainties, but only if the faults are buried and have lengths much shorter than if they marked a plate boundary.

Both the Bhuj-Kachch and New Madrid earthquakes may yet be driven by stresses propagated from far-field plate boundaries or by local generation. Whether the two sets of earthquakes are tectonic analogs is still open for debate. The answer has consequences for seismic hazard beyond the confines of each of these earthquakes. For example, if plate compression is the fundamental driving force behind the Bhuj-Kachchh earthquakes, then these earthquakes are accommodating displacement along an Euler-latitude, such that elsewhere on the same latitude, seismic hazard should be concomitantly lowered. Molnar (1987) made a similar suggestion in reference to the 1897 Assam earthquake, which sits in front of the Bhutan Himalaya and partly fills in the gap between the 1934 and 1950 Himalayan earthquakes.

6.5 DISCUSSION

Can the Bhuj earthquake outline a research agenda for seismic hazard analyses within the interior of plates?

Two features of the Bhuj earthquake provide important clues in trying to unravel the enigmatic New Madrid earthquakes. The first is the clear anomaly in scaling, leading to a stress drop that is possibly twice and maybe three times as large as typical earthquakes of this size. The second is the depth extent of rupture. Judging from the aftershocks, the Bhuj earthquake appears to have ruptured the entire crust. Prior to the Bhuj earthquake, the depth extent of large earthquakes in the New Madrid region was a contentious issue, because it was a critical source of seismic moment energy that would help explain the vast area of strong ground motion. Conventional wisdom states that, for non-subduction earthquakes, the largest upper crustal earthquakes are bound by the base of the upper crust, and that the lower crust flows aseismically (e.g., Scholz, 1990). Most of this wisdom, unfortunately, comes from plate boundaries and particularly from California. Scaling of rupture parameters in these sorts of environments is dictated in essence by the infinitely long dimension of active faults (Bodin and Brune, 1996). The Bhuj aftershocks and present slip-inversion models (as well as the recent analysis of the 1897 Assam earthquake by Bilham and England (2001)) give credence to the notion that the 1811-1812 New Madrid ruptures may well have ruptured much of the crust.

These observations suggest some important questions for future research. In particular, it is important to know if the normal state of background seismicity in the Bhuj region is similar to that in New Madrid. That is, does background seismicity occupy only the upper crust or do earthquakes occur within the lower crust on a normal basis? An important corollary but one that the author does not address here is: do aftershocks decay in the same fashion (following Omori's Law with the same coefficients) as the standard California model, or is there a significant difference in the decay behaviour? This is critical knowledge for seismic hazard, because aftershocks are an immediate impediment to recovery operations and they cause considerable damage to an already damaged infrastructure. And thirdly, do earthquakes that occur on the nearby plate boundary (e.g., Pakistan and to the north along the Nepalese Himalaya) yield similarly large strong ground shaking patterns in the Indian subcontinent? In other words, how important is it to the distribution of ground motion that the earthquake be located away from the boundary?

Of the three characteristics identified earlier (rupture, path, and tectonic), the Bhuj the and New Madrid earthquakes qualify well as analog earthquakes in the first two of these. The most compelling probable similarities of scaling and depth of rupture set the basic template for distributing strong ground motions. On top of this, quite literally, each region shares the common feature of having a relatively thin veneer of sedimentary rocks on top of an old and strong basement. These similarities alone make the Bhuj earthquake an invaluable resource for anyone seriously interested in seismic hazard in the central USA and the Indian subcontinent.

How these earthquakes are driven remains unknown, but it is clear that for each case, the evaluation of driving forces requires high-quality geodetic data that goes hand-in-hand with incrementally evolving tectonic models.

Finally, I have tried to avoid using the language of intraplate and interplate earthquakes, because such labels tend to carry a lot of baggage. There are at least three characteristics of earthquakes that can be described in terms of analog behaviour: the rupture process, the material through which the seismic waves propagate, and the tectonic driving force. Not all earthquakes within plates will share each of these characteristics, and nor will all earthquakes occur within a plate margin. Earthquakes and their associated hazard are complicated things. The use of jargon can be convenient for communication, but it also tends to fossilize our original meaning and ultimately ends up causing more confusion than convenience.

ACKNOWLEDGEMENTS

I am very grateful to my colleagues at the Center for Earthquake and Information, University of Memphis, for many interesting discussions. I thank Arch Johnston and Eugene Schweig in particular for being generous with their time and ideas. I also thank Mian Liu for sharing a preprint of the Li et al., manuscript. Support by the National Science Foundation,

(NSF EAR 9803484) is gratefully acknowledged. This is CERI contribution 445.

REFERENCES

Antolik, M. and Dreger, D. 2001. Source rupture processs of the 26 January, 2001, Bhuj India, earthquake, (M 7.6), EOS, Trans. AGU 82(47), Fall Meeting Suppl., Abstract S52G-03, 2001.

Bendick, R. and R. Bilham. 1999. Search for buckling of the southwest Indian coast related to Himalayan collision. In Himalaya and Tibet: mountain roots to mountain tops. Ed. A. Macfarlane, R. Sorkhabi, and J. Quade. Geological Society of America Special Paper **328**. pp. 313-323

Bendick, R., Bilham, R., Fielding, S. E., Gaur, V., Hough, S. E., Kier, G., Kulkarni, M. N., Martin, S., Mueller, K., and Mukul, M., 2001. The January 26, 2001 Bhuj earthquake, India, Seism. Res. Letters, 72, 328-335.

Baker, W. E., 184. Remarks on the Allah Bund and on the drainage of the eastern part of the Sind basin, Trans. Bombay Geogr. Soc., 7, 186-188.

Bernard, M., Shen-Tu, B., Holt, W. E., and Davis, D. E. 2000. Kinematics of active deformation in the Sulaiman Lobe and Range, Pakistan, J. Geophys. Res., 105, 13,253-13,279.

Bilham, R. 1998. Slip parameters for the Rann of Kachchh, India, 16 June 1819 earthquake quantified from contemporary accounts, in Stewart, I. S. & Vita-Finzi, C. (Eds) Coastal Tectonics. *Geological Society London*, **146,** 295-318.

Bilham, R. and England P., 2001. Plateau pop-up during the great 1897 Assam earthquake, Nature, 410, 806-809.

Bilham, R. and Gaur, V.K., 2000 Geodetic contributions to the study of the seismotectonics in India, Current Science, 79, 1259-1269.

Biswas, S. K., 1982 Rift basins in the western margin of India and their hydrocarbon prospects, Bull. Am. Assoc. Pet. Geol., 66, 1497-1513.

Biswas, S. K., 1987 Regional tectonic framework, structure and evolution of the western marginal basins of India, Tectonophysics, 135, 307-327.

Bodin, P., and Brune, J.N., 1996. ÒOn the Scaling of Slip with Rupture Length for Shallow Strike-Slip Earthquakes: Quasi-static Models and Dynamic Rupture PropagationÓ. Bull. Seis. Soc. Am., 86, p. 1292-1299.

Chiu, J. M., Johnston, A. C., Yang, Y. T. 1992 Imaging the active faults of the central New Madrid seismic zone using Panda array data, Seismological Research Letters, 63, 375-393.

DeMets C., Gordon, R. G, Argus, D. F., and Stein, S., 1990 Current plate motions, Geophys. J. Int., 101, 425-478.

Ellis, M., Schweig, E., Prejean, S. G., and Johnston, A. J. 1995. Rupture scenario for the 1811-1812 New Madrid earthquakes based on static stress relaxation and change of topography. EOS, Transactions, American Geophysical Union, v.76, p F381.

Ellis, M., Gomberg, J., and Schweig, E. S. 2001. Indian earthquake may serve as an analog for New Madrid earthquakes, EOS, 82, 345 350.

Ervin, C. P. and McGinnis, L. D. 1975. Reelfoot rift: reactivated precurser to the Mississippi Embayment, Geol. Soc. Am., Bull., 86, 1287-95.

Eshleby, J. 1957 The determination of the elastic field of an ellipsoidal inclusion and related problems, Proc. Royal Soc. Lond., Series A, 241, 376-96.

Gomberg, J. and Ellis, M., 1994 Topography and tectonics of the central New Madrid seismic zone: Results of numerical experiments using a three-dimensional boundary element program, J. Geophys. Res., 99, 20299-20310.

Grana, J.P. and Richardson, R.M., 1996. Tectonic stress within the New Madrid seismic zone, J. Geophys. Res., 101, 5445-5458.

Grollimund, B. and Zoback, M. D. 2001. Did deglaciation trigger intraplate seismicity in the New Madrid seismic zone?, Geology, 29, 175-178.

Hanks, T. and Johnston, A.C., 1992. Common features of the excitation and propagation of strong ground motion for North American earthquakes, Bull. Seismol. Soc. Am., 82, 1-23.

Haq, S. S. B. and Davis, D. M. 1997. Oblique convergence and the lobate mountains belts of western Pakistan, Geology, 25, 23-26.

Hildenbrand, T. G. 1985. Rift structures from the northern Mississipi embayment from the analysis of gravity and magnetic data, J. Geophys. Res., 90, 12,607-22.

Holt, W. E. and five others. Velocity field in Asia inferred from Quaternary fault slip rates and global positioning system observations, J. Geophys. Res.,105, 19,185-19, 209.

Hough, S.E., Armbruster, J.G., Seeber, L., and Hough, J.F., 2000. On the modified Mercalli intensities and magnitudes of the 1811-1812 New Madrid earthquakes, Journal of Geophysical Research, 23,839-23,864.

Hough, S., Martin, S., Bilham, R., and Atkinson, G.2001. The January 26, 2001 Mw7.6 Bhuj, India, earthquake: observed and predicted ground motions, EOS, Trans. AGU 82(47), Fall Meeting Suppl., Abstract S52G-02.

Johnston, A.C. 1996 Seismic moment assessment of earthquakes in stable continental regions – III. New Madrid 1811-1812, Charleston 1886, and Lisbon 1755, Geophys. J. Int., 126, 314-344.

Johnston, A.C. and Schweig, E. S. 1996. The enigma of the New Madrid earthquakes of 1811-1812, Annual Rev. Earth Planet. Sci., 24, 339-384.

Kelson, K. I., Van Arsdale, R. B., Simpson, G. D., and Lettis, W. R. Assessment of the style and timing of late Holocene surficial deformation along the central Reelfoot scarp, Lake County, Tennessee, Seismological Research Letters, 63, 349-356.

Kenner, S.J. and Segall, P. 2000. A mechanical model for intraplate earthquakes: Application to the New Madrid seismic zone, Science, 289, 2329-2332.

Kumar, P., Tewari, H. C., and Khandekar, G. 2000. An anomalous high velocity layer at shallow crustal depths in the Narmada zone, India, Geophys. J. Int., 142, 95-107.

Langslon, C. A. 2001 Teleseismic source models and source depth constraints for events of the Bhuj earthquake sequence, EOS, Trans. AGU 82(47), Fall Meeting Suppl., Abstract S52G-05, 2001.

Li, Q., Liu, M., and Yang, Y., 2002. The 01/26/2001 Bhuj, India earthquake: intraplate or intraplate ? In: Plate Boundary zones (Stein and Freymueller eds.), Am. Geoghys. Union Geodynamics Ser. 30, DOI: 10/1029/030 GD16.

Molnar, P. 1987. The distribution of intensity associated with the great 1897 Assam earthquake and bounds on the extent of the rupture zone, J. Geol. Soc. India, 30, 13-27.

Mori, J., Sato, T., and Negishi, H. 2001. Slip distribution of the 2001 west India earthquake, EOS, Trans. AGU 82(47), Fall Meeting Suppl., Abstract S51B-0604.

Negishi, H. and 5 others 2001 Aftershock distribution of the 2001 Gujarat, India earthquake (Mw 7.7) from temporary field observations: small and deep orientation of the fault plane, EOS, Trans. AGU 82(47), Fall Meeting Suppl., Abstract S51B-0606.

Newman, A., Stein, S., Weber, J., Engeln, J., Mao, A., and Dixon, T. Slow deformation and lower seismic hazard at the New Madrid seismic zone, 284, 619-621.

Newmann, A., Stein, S., Weber, J., Engeln, J., Mao, A., and Dixon, T. 1999. Slow Deformation and Lower Seismic hazard at the New Madrid Seismic Zone, Science, 284, 619-622.

Newman, Schneider, A.J., Stein, S., and Mendez, A. 2001. Uncertainties in seismic hazard maps for the NMSZ and implications for seismic hazard communications, SRL, 72, 647-661.

Oldham, R. D. 1898 A note on the Allah Bund in the Rann of Cutch, Mem. Geol Survey of India, 28, 27-30.

Paul, J., Burgmann, R., Gaur, V.K., Bilhm, R., Larson, K.M., Ananda, M.B., Jade, S., Mukal, M., Anupama, T.S., Satyal, G., and Kumar, D., 2001. The motion and active deformation of India, Geophys. Res. Letts., 28, 647-650.

Pollitz, F.F., Kellog, L., Bürgmann, R. 2001. Sinking Mafic Body in a Reactivated Lower Crust: A Mechanism for Stress Concentration at the New Madrid Seismic Zone, Bull. Seismol. Soc. Am., 91, 6, 1822-1897.

Rajendran, C.P., and Rajendran, K. 2001. Characteristics of deformation and past seismicity associated with the 1819 Kutch earthquake, northwestern India, Bull. Seism. Soc. Am., 91, 407-426.

Raphael, A. J., Bodin, P., Horton, S., and Gomberg, J. 2001 Preliminary double-difference relocations of Bhuj aftershocks, EOS, Trans. AGU 82(47), Fall Meeting Suppl., Abstract S51B-0605.

Richardson, R. M. 1992. Ridge forces, absolute plate motions, and the intraplate stress field, J. Geophys, Res., 97, 11739-11748.

Russ, D. P. 1979. Late Holocene faulting and earthquake recurrence in the Reelfoot Lake area, northwestern Tennessee, Geol. Soc. Am., Bull., Part I, 90, 1013-18.

Scholz, C. 1990. The Mechanics of Earthquake Faulting, Cambridge University Press, Cambridge, p439.

Sieh, K. and 19 others. 1993. Near-field investigations of the Landers earthquake sequence, April to July, 1992, Science, 260, 171-176.

Stein, S., Okal, E., Sella, G., and Schoonover, M., 2001. The Bhuj earthquake and tectonics of the Indian plate's diffuse western boundary, EOS, Trans. AGU 82(47), Fall Meeting Suppl., Abstract S52G-07.

Stuart, W.D., Hildenbrand, T.G., and Simpson, R.W. 1997 Stressing of the New Madrid seismic zone by an lower crust detachment fault, J. Geophys. Res.,102, 27, 27.623-27,633.

Tuttle, M., Chester, J., Lafferty, R., Dyer-Williams, K., and Cande, R. 1999 Paleoseismology study of northwest of the New Madrid seismic zone, U.S. Nuclear Reg. Comm. Final Rept. NUREG/CR-5730, p.137.

Tuttle, M. P., Johnston, A., Rajendran, C.P., Rajendran, K., Thakkar, M., 2001. Liquefaction features induced by the Republic Day earthquake and comparison with features related to the 1811-1812 New Madrid earthquakes, American Geophysical Union, Transactions, EOS. v. 82, p.261.

Wells and Coppersmith, Bull. 1994 Seism. Soc. Am., 84, 974-1002.

Wesnousky, S. G. and six others 2001. Eight days in Bhuj: Field report bearing on surface rupture and genesis of the 26 January 2001 earthquake in India, Seism. Research Lett., 72, 514-523.

Wu, P. and Johnston, P. 2000. Can deglaciation trigger earthquakes in N. America, Geophys. Res. Letts., 27, 1323-1326.

Zoback, M.L. 1992 Stress field constraints on intraplate seismicity in eastern North America, J. Geophys. Res., 97, 11761-11782, 1992.

7

Seasonality Bias of Reservoir-Induced Seismic Activity in the Koyna Area, India

Rama Krishna Tiwari, Sabrina Leonardi and Hans-Joachim Kümpel

ABSTRACT

We have carried out statistical analyses to investigate the possible seasonality bias of seismic activity in the Koyna region, India. The study is based on all earthquakes with $M \geq 4.0$ from March 1968 till April 1996. Application of the Schuster test reveals that these earthquakes originate from a non-random process possibly associated with deterministic seasonality bias. We assessed the dimensionality of earthquake generating mechanisms by non-linear predictability analysis on phase portrait constructed by recurrence time series. The result of non-linear forecasting analyses suggest that the earthquake process in the Koyna region evolves on a high dimensional chaotic system. The seismically active part of the Koyna area appears to have come in the state of 'criticality' because of the impounding of the reservoir. The quasi-cyclic loading of the reservoir, due to seasonal rainfall, leads to enhanced seismic activity after a certain delay in time. Assessments based on average crustal parameters indicate that the time needed for pore fluid pressure to propagate downward to the seismogenic depths in the Koyna region could be approximately one to three months, most likely during the months of October to December. We hypothesize that stochastic seasonal resonance combined with noise affect pore pressure variation and cause subsequent triggering of self-organized earthquakes. Evidence for high dimensional chaos associated with seasonal bias in the Koyna region may provide useful constraints for testing models and criteria to assess earthquake hazards on a more rigorous and quantitative basis.

7.1 INTRODUCTION

The mechanism of reservoir-induced seismicity (RIS) is a widely debated phenomenon (e.g. Allen, 1982). Several lines of evidence suggest that changes in pore fluid pressure, due to increase in water level in a reservoir, is one of the main causes of triggering RIS. Some researchers believe that earthquakes occurring around reservoirs are due to intraplate activity, i.e. the coincidence in space and in time is just by chance, while others have proposed that triggering of earthquakes may be phreatically caused by loading through the reservoir and/or by pore pressure diffusion into the ground (Bell and Nur, 1978; Talwani and Acree, 1984; Simpson, 1986; Chadha et al., 1999; Rai et al., 1999). Gupta et al., (1972) have identified several important factors, such as increase of water level, duration of loading, reduced maximum level, and duration of retention of high water level as possibly affecting the frequency and magnitude of earthquakes near artificial reservoirs.

During the past few years efforts have been made to study the nature of the dynamics of reservoir induced seismicity quantitatively. Several researchers considered rainfall as the main source of feeding reservoirs and aquifers and there have been several attempts to investigate the interrelations between rainfall and seismic activity (e.g. Huang et al., 1979; Mil'kis, 1986; Costain et al., 1987; Kafri and Shapira, 1990; Albarello et al., 1991; Muco, 1995; 1999; Tiwari and Rao, 2001). Increase of water level in reservoirs is generally caused by seasonal rainfall, which, in case of the prominent Koyna region in Maharastra, India, approximately lasts from June to September. We consider rainfall as one of the main sources of feeding a reservoir; seasonality could be reflected in the earthquake data, if it is pronounced in the hydrological year.

An attempt has also been made to characterize the nature of dimensionality (Srivastava et al., 1994) using the non-linear Grassberger-Procaccia (G-P) algorithm (Grassberger and Procaccia, 1983). It may, however, be noted that identification of chaotic dynamics using the (G-P) algorithm may encounter at least two main problems: (1) Detecting a chaotic behaviour using the above algorithm requires a large number of data points which is often not available. (2) Even if enough data are available a finite attractor dimension may not be actually indicative of deterministic chaos due to low signal-to-noise ratio (Osborne and Provenzale, 1989). In such a situation the resulting anomalous scaling behaviour could be simply a hallmark of random fractal. Such fractal characteristics have their origin in re-scaled random processes generally known as fractional Brownian motion and are highly unpredictable. Fractals from the cellular automata seismicity models (dynamically high dimensional and stochastic in nature) fall into this class. Distinguishing these classes of evolutionary seismicity is essential for understanding and constraining models of crustal dynamics.

Induced earthquake activity in the Koyna region is highly prone to low signal-to-noise ratio. It is expected that any deterministic or orderly behaviour may be contaminated by noisy environment. Hence, any attempt to visualize or understand the nature of dynamics using qualitative or non-linear analyses using the G-P algorithm will be simply misleading. A new study is, therefore, undertaken here, in which the largest Koyna region earthquakes from March 1968, three months after the devastating M 6.3 event in December 1967, till April 1996 are utilized. We address here two main problems: (1) Search for some order, if any, related to the seasonal bias in these data using the well known Schuster test of statistical significance, hitherto not reported in any of the previous studies. The question of seasonality in the occurrence of the Koyna region earthquakes is important, since a statistically significant average recurrence time would provide some constraints for predictability of RIS in this region. (2) Application of a modern non-linear forecasting technique to characterize the nature of the dynamics (e.g. harmonic, stochastic, non-linear, chaotic, or random). Quantification of the nature of the dynamics allows one to better understand the dynamical conditions prevailing in the region.

7.2 THE DATA

The Koyna region is situated in the Peninsular Shield of India, which had been long considered relatively free from any significant seismic activity. After the M 6.3 earthquake of December 10, 1967, seismicity has continued in the Koyna area, where earthquakes with M ≥ 4.0 are not rare (Gupta, 1992). A new catalogue recently updated from records of the National Geophysical Research Institute (NGRI), Hyderabad, is utilized here. The total data enlists 98 events of magnitude equal to or exceeding 4.0 that occurred between March 1968 and April 1996 (H.K. Gupta, personal communication). The magnitude of each earthquake is based on the estimate of its coda duration, and only events with signal amplitudes above a preset signal-to-noise ratio are considered here. A list of the 30 events with M ≥ 4.5 is given in Table 7.1. Fig. 7.1 shows month- and year-wise distribution of these data. It is evident that about 63% of the events occurred during the months from July to December.

Table 7.1:   *List of earthquakes with magnitude M ≥ 4.5 that occurred in the Koyna region from March 1968 to April 1996. ω_i represents the phase angle measured in days from the beginning of the year.*

Event	Year	m	d	ω_i	M
1	1968	03	04	64	4.6
2	1968	08	31	244	4.5
3	1968	10	29	303	5.0
4	1969	06	27	178	4.6
5	1970	05	27	147	4.7
6	1970	09	25	268	4.6
7	1973	10	17	290	5.0
8	1974	07	29	210	4.7
9	1976	03	14	74	4.7
10	1976	12	12	347	4.5
11	1977	09	19	262	4.5
12	1978	12	12	346	4.5
13	1980	02	06	37	4.7
14	1980	09	02	246	4.9
15	1980	09	20	264	5.0
16	1980	10	03	277	4.6
17	1968	10	04	278	4.8
18	1982	04	25	115	4.7
19	1983	02	05	36	4.7
20	1983	09	25	268	4.9
21	1984	09	25	268	4.6
22	1985	11	15	319	4.5
23	1989	10	29	302	4.5
24	1991	06	01	152	4.7
25	1993	08	28	240	4.9
26	1993	09	03	246	4.8

(Table. 7.1 contd.)

(Table. 7.1 contd.)

27	1993	12	08	342	5.1
28	1994	02	01	32	5.2
29	1995	03	12	71	4.9
30	1996	04	26	117	4.5

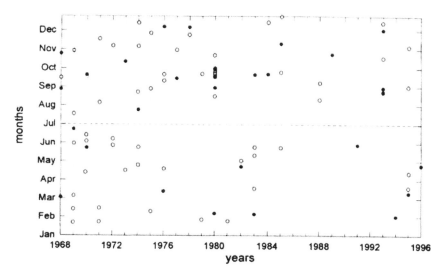

Fig. 7.1　Month- and year-wise distribution of earthquakes with M ≥ 4.0 that occurred in the Koyna region from March 1968 to April 1996. Black circles indicate earthquakes with M ≥ 4.5; white circles those with 4.0 ≤ M ≤ 4.5. Abscissa has events projected to the year.

7.3　METHODS OF ANALYSIS

7.3.1　Search for Order Using the Schuster Statistical Test

To see some pattern in the Koyna earthquakes data we applied a well-known methodology originally devised by Schuster (1897). It has also been applied by several authors in different contexts (e.g. Heaton, 1975; McClellan, 1984; Stothers, 1989; Fadeli et al., 1991; Muco, 1999). The Schuster test quantifies the tendency of events to occur around a certain phase of a given cycle. Using the phase angle ω_i (measured from the beginning of the cycle) for each date t_i of occurrence, the Schuster test proceeds by adding all N phases vectorially, and so finds a resultant vector

of phase angle $\phi = \arctan B/A$ and length $R = \sqrt{A^2 + B^2}$, where $A = \Sigma \cos \omega_i$ and $B = \Sigma \sin \omega_i$. A vector longer than R will occur by chance from a set of N completely random phases with a probability $P = e^{-R2/N}$, for $N > 8$. In computing ω_i here, the year is taken to have 365 days and February 29 is treated as March 1.

The application of the Schuster test to the dates of the Koyna region earthquakes with M ≥ 4.0 yields a high probability that these earthquakes come from a source of non-random distribution. The results show that the null hypothesis of randomly distributed events can be rejected at a significance level P smaller than 0.05, specifically 0.002 for dates (angle ϕ) around October, 12 (Table 7.2). In order to check any accidental significance that could result from non-stationarity of the data, we further analysed the total data set using a moving window of 30 events width, successively shifted by 10 events. In most cases, the statistical significance is 90% or higher (Table 7.2). In the second step, we selected only 30 events with magnitude M ≥ 4.5. As before, one analysis has been performed for the total data set, two more for moving windows of widths 20 and 25 events, respectively. The results are noted in Table 7.3. Again, a statistically significant clustering is observed for the dates in the month of October (Fig. 7.2), indicating that the probability for the seismicity in the Koyna area to have occurred randomly in this month is lowest.

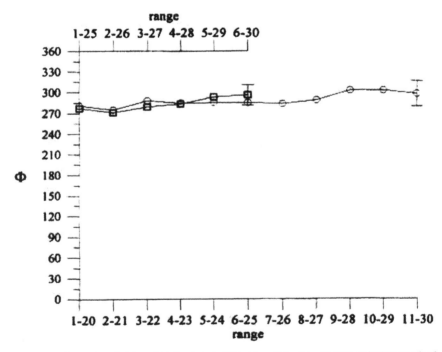

Fig. 7.2 Clustering of dates of highest probability of earthquake occurrence in the month of October (data from Table 7.3); error bars represent Δφ values.

Table 7.2 *Results of the Schuster test for the 98 earthquakes with M ≥ 4.0 that occurred in the Koyna region from March 1968 to April 1996. 30 events width-moving window with 10 events step was applied. The phase angle Φ is already converted into a date (month, day) of highest probability; ΔΦ denotes stochastic resolution.*

Range	R	P	φ		Δφ
1-98	24.87	0.002	10	12	3.7
1-30	3.66	0.6	2	5	12.7
11-40	3.44	0.7	2	23	12.7
21-50	7.79	0.1	11	5	12.7
31-60	15.23	0.0004	10	22	12.7
41-70	18.11	0.00002	10	22	12.7
51-80	12.14	0.007	10	9	12.7
61-90	10.62	0.02	10	3	12.7
71-98	4.95	0.4	10	7	12.7

Table 7.3 *Results of the Schuster test for the earthquakes of Table 7.1. Moving windows of 25 events width and of 20 events width were applied. Φ and ΔΦ as in Table 7.2.*

Range	R	P	φ		Δφ
1-30	8.07	0.1	10	13	12.2
1-25	8.95	0.04	10	4	14.6
2-26	10.7	0.01	9	28	14.6
3-27	10.23	0.01	10	6	14.6
4-28	8.94	0.04	10	10	14.6
5-29	8.46	0.06	10	20	14.6
6-30	8.29	0.06	10	23	14.6
1-20	6.09	0.2	10	8	18.3
2-21	7.91	0.04	10	2	18.3
3-22	7.87	0.05	10	15	18.3
4-23	7.87	0.04	10	11	18.3
5-24	7.48	0.06	10	12	18.3
6-25	8.93	0.02	10	12	18.3
7-26	8.76	0.02	10	10	18.3
8-27	8.31	0.03	10	15	18.3
9-28	8.09	0.04	10	29	18.3
10-29	8.13	0.04	10	29	18.3
11-30	6.46	0.1	10	24	18.3

Accordingly, seasonality appears to play a vital role for triggering earthquake activity in this region. Clearly, seasonal rainfall and reservoir filling add some load to the surface and increase the pore fluid pressure due to the decrease in pore volume, caused by compaction (Bell and Nur, 1978; Gupta, 1992).

7.3.2 Possible Explanation of the Statistical Test Result

Several model studies have also stressed the significant role played by pore pressure diffusion due to fluid migration reducing rock strength in saturated formations and thereby inducing earthquakes (Bell and Nur, 1978; Talwani and Acree, 1984; Simpson, 1986). Lee and Wolf (1998) suggest that the time t, required for a surface pore pressure disturbance to propagate downward to a depth z is given by

$$t \cong 7.9 \times 10^{-22} \, z^2/k \tag{1}$$

Herein, time t is expressed in years and k is permeability, generally observed to range from 10^{-18} to 10^{-12} m^2 in deep, fractured, brittle crust (Clauser, 1992; Lee and Wolf, 1998). For the period studied, 80% of the earthquakes in the Koyna area concentrate at hypocentral depths from 4 to 10 km, 15% from 1 to 4 km and 5% at about 12 km (Prantik Mandal, personal communication). The results of the present analysis suggest that it takes approximately one month after maximum loading of the reservoir, or three months after highest rate of refilling, to trigger most of the earthquakes. Using e.g., a permeability $k = 10^{-13}$ m^2 and a source depth $z = 6$ km, the above model provides approximately three months time for a surface disturbance to trigger most of the earthquakes at that depth. Fig. 7.3 displays different scenarios for other sets of parameter values. It is seen that the most probable dates of increased seismic activity depend upon effective hydraulic connectivity in the sub-surface, which is *de facto* unknown. This analysis, however, cannot discriminate between two possible sources for seasonality in the seismic activity: reservoir related, where the source is restricted to the reservoir itself and the pore pressure increase is equivalent to 35 m height in water level, on average, or rainfall related which is of less amplitude in pore pressure increase but more widespread than the reservoir effect. Note also that delays due to propagation of pore pressure disturbances over lateral distances are ignored.

7.3.3 Order in Midst of Chaos, A Non-linear Forecasting Approach

The earthquake generating process is generally known to be non-linear. Fig. 7.4 shows recurrence time series (time interval between successive earthquakes) for the Koyna region earthquakes.

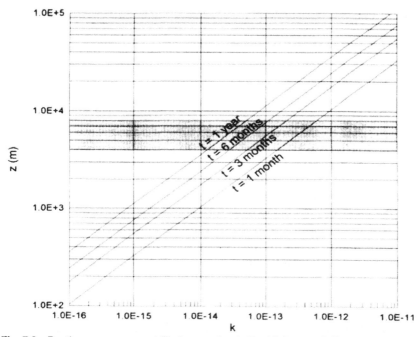

Fig. 7.3 Depth z versus permeability k according to Eq. (1) (see text). Straight lines represent correspondent triggering times of earthquakes. The grey box gives typical z and k values for the Koyna area.

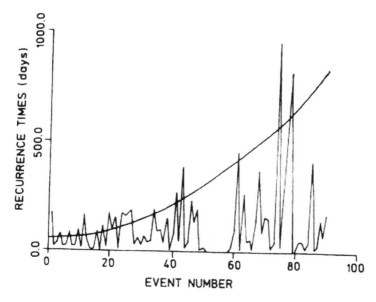

Fig. 7.4 Recurrence (time interval between successive earthquakes, TIBSE) time series for the Koyna region earthquakes (M ≥ 4). Horizontal scale represents event numbers and vertical scale recurrence time interval.

From Fig. 7.4 it is evident that the recurrence time series exhibits a non-uniform recurrence interval and can be approximated by a non-linear process. In a non-linear system a quasi-cyclic pattern may exist, but its continuity and/or uniformity throughout the total record cannot be guaranteed. Also, in such a non-linear dynamical system, it is not possible to specify the position of such pattern precisely due to sensitive dependence on the initial conditions. Hence, the predicted phase space position will become increasingly inaccurate as time grows (as the length of prediction increases). To characterize the nature of dynamics from such a non-linear data set, we here employ a non-linear forecasting technique originally devised by Sugihara and May (1990). This method of forecasting a future value is an extension of the zero-order technique of Farmer and Siderowich (1987). Several workers (Tsonis and Elsner, 1992; McCloskey, 1993; Tiwari and Rao, 1999) have already demonstrated the applicability of this technique to multidimensional data to distinguish chaos from stochastic/random fractals. The method of prediction is briefly described below.

(a) Method forecasting analysis: First a subset also called 'simplex' is extracted from the complete data set. The 'simplex' is the smallest 'm-value' that contains a vector assigned to the actual one in the m-space. The usual procedure is to transform the data series in 'm' dimensional state vectors X(t) by assigning coordinates (Takens, 1981). Here, we use available earthquake time series to construct the phase space using the appropriate time delay-embedding dimension. Accordingly, an m-component state vector X_i from a time series x(t) can be given as:

$$X_i = \{x_1(t_i)\,, x_2(t_i), \ldots \ldots \ldots \ldots ., x_m(t_i)\} \tag{2}$$

where $x_k(t_i) = x[t_i + (k - 1)\tau]$ and τ is an appropriate delay time. The m + 1 nearest vector defines the vertices of the simplex (Fowler and Roach, 1993). Sugihara and May (1990) make predictions by giving an expected weight to the shortest distances between the vector assigned to the actual one as m + 1 vertices. The method is robust and works well on even shorter data thus overcoming the earlier difficulties where huge amount of data were required to carry out non-linear analyses using the G-P algorithm. In essence, the non-linear forecasting approach predicts the position to which a point in the plane moves by observing the behaviour of a 'simplex' of its nearest neighbours. The criterion for choosing the 'simplex' relies on sorting all of the points in the phase portrait in order to obtain their absolute distance d from the central points (McCloskey et al., 1991) where

$$d = \left[\sum_{j=1}^{m} (X_{cj} - X_{ij})^2 \right]^{1/2} \qquad (3)$$

Herein, X_{cj} is the j-th coordinate of the central point and x_{ij} is the j-th coordinate for the point on the phase portrait.

The optimum 'simplex' is the one for which the product of phase volume V and total vector displacement d from the central point is a minimum. In other words, the optimal 'simplex' of all available simplexes is the one, which has minimum volume, where V is given by

$$V \propto 1/(m+1) \left\{ \sum_{i=1}^{m+1} \left[\sum_{j=1}^{m} (X_{cj} - X_{ij})^2 \right]^{1/2} \right\}^m \qquad (4)$$

and

$$d = \left\{ \sum_{i=1}^{m+1} \left[\sum_{j=1}^{m} \Phi (X_{cj} - X_{ij})^2 \right]^{1/2} \right\}^m \qquad (5)$$

and where $\Phi = 1$ or $x_{cj} - x_{ij} > 0$ and $\Phi = -1$ for $x_{cj} - x_{ij} < 0$.

The behaviour of any measure of the agreement between predictor and true data (e.g. the correlation coefficient, r(t)) gives an indication of the nature of dynamics. To ensure the stability and hence accurate predictability of the nature of dynamics, a proper choice of embedding dimension 'm' is essential, i.e. m > D is the minimum requirement where D is the attractor dimension. Figure 7.5 shows the correlation coefficient between actual and predicted values for a one-step prediction, r(1), for the actual time series as a function of the embedding dimension. In Fig. 7.5, a sudden increase of r(t) for an embedding dimension of 2-3 indicates that the underlying earthquake dynamics is non-random. A high correlation found for embedding dimension of 2-3 can be used for further study (Tsonis and Elsner, 1992).

(b) Predictive correlation analysis: After computing the predicted values for the Koyna region and three other data sets (e.g. random white noise, stochastic, and chaotic) using the technique described, we define the Pearson's correlation coefficient r(t) between the actual and the predicted values as

$$r(t) = \Sigma[(X_t - X_m)(Y_t - Y_m)]/\sqrt{[S(X_t - X_m)^2 \, S(X_t - X_m)^2]]} \qquad (6)$$

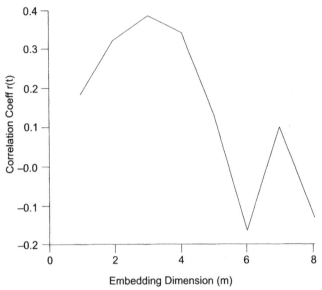

Fig. 7.5 A plot of correlation coefficients between actual TIBSE and corresponding predicted values (vertical scale) versus embedding dimension (horizontal scale).

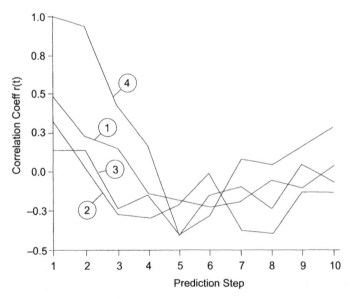

Fig. 7.6 Correlation coefficients between actual TIBSE and corresponding predicted values (vertical scale) versus prediction time scale for 4 different models: 1. Actual Koyna earthquake data, 2. Stochastic data, 3. Random data, 4. Chaotic data (see text).

where X_t and Y_t are the actual and the predicted values and X_m and Y_m are the mean of the actual and the predicted values, respectively.

Curves 1 to 4 in Fig. 7.6 show the comparison of the predictability correlation coefficient r(t) between actual and predicted data, using the algorithm described in earlier section, as a function of prediction interval T_p for the four different data sets. These data are: Koyna data, data generated using the autoregressive model, random models, chaotic time series generated using the logistic model. It is known that a short-term predictability is a characteristic feature of chaotic data. The predictability of a chaotic signal is, therefore, expected to be fairly good with high initial correlation coefficient for short time interval apparently evident in curve 4 of Fig. 7.6. In contrast, the same is not true for random data (curve 3) as these are never predictable and show almost zero level correlation.

Contrary to these, both stochastic data with expected fixed red noise (curve 2) and the actual Koyna earthquake data (curve 1) compare well and exhibit a low level of correlation which is indicative of random fractal (similar to Brownian noise). Such a high dimensional fractal originates from 'self organized' critical processes and have been reported widely in many earthquake processes (Bak and Tang, 1989).

7.4 CONCLUSIONS

Based on the Schuster statistical test we conclude that (1) the Koyna earthquakes come from a non-random process and (2) the most likely month of enhanced earthquake activity appears to be October, i.e. approximately one month after maximum loading and three months after highest rate of refilling in the area. The most probable dates of anomalous earthquake activity should strongly depend on the hydraulic connectivity in the region, which is not known. (3) Based on non-linear forecasting results, we conclude that there is evidence for a mixed response of high dimensional chaos and a stochastic nature of cyclicity in the Koyna region earthquakes. The seismically active part of the Koyna area can be understood as being in a state of 'criticality' due to initial impounding of the reservoir. Since then, the change in pore fluid pressure, due to seasonal rainfall and/or filling of the Koyna Dam reservoir, seems to be the most likely reason for triggering of earthquakes in the Koyna area.

ACKNOWLEDGEMENTS

We are grateful to Dr. H.K. Gupta for providing the data and insights. One of the authors (R.K.Tiwari) acknowledges the support from the Deutsche

Akademische Austauschdienst, Bonn, for visit to the University of Bonn. We are also thankful to Shri K.N.N. Rao for his help in computations and to Shri V. Subrahmanyam and Ms. S. Sri Lakshmi for their assistance in the preparation of the manuscript.

REFERENCES

Albarello, D., Ferrari, G., Martinelli, G., and Mucciarelli, M. 1991. Well-level variation as possible seismic precursor: a statistical assessment from Italian historical data. Tectonophysics, 193, 385-395.

Allen, C.R. 1982. Reservoir induced earthquakes and engineering policy. Calif. Geol., Nov. 1982, 248-250.

Bak, P., and Tang, C. 1989. Earthquake as a self organized critical phenomenon. J. Geophys. Res., 94 (B11), 15635-15637.

Bell, M.L., and Nur, A. 1978. Strength changes due to reservoir-induced pore pressure and application to Lake Oroville. J. Geophys. Res., 83, 4469-4483.

Chadha, R.K., Rastogi, B.K., Mandal, P., and Sarma, C.P.S. 1999. Reservoir associated seismicity (RAS) in Indian Shield. Mem. Geol. Soc. India, 43, 415-423.

Clauser, C. 1992. Permeability of crystalline rocks. EOS Trans. AGU, 73, 233, 237-238.

Costain, J.K., Bollinger, G.A., and Speer, J.A. 1987. Hydroseismicity: a hypothesis for the role of water in the generation of intraplate seismicity. Geophys. Res. Lett., 58(3), 41-64.

Fadeli, A., Rydelek, P.A., Emter, D., and Zürn, W. 1991. On volcanic shocks at Merapi and tidal triggering. In: Volcanic Tremor and Magma Flow (R. Schick & R. Mugiono, eds.), Forschungszentrum Jülich, Sci. Service Int. Bureau, 165-181.

Farmer, J.D., and Siderowich, J.J. 1987. Predicting chaotic time series. Phys. Rev. Lett., 59, 845-848.

Fowler, A.D., and Roach, D.E. 1993. Dimensionality analysis of time series data: Nonlinear method. Computer and Geosciences, 19(1), 41-52.

Grassberger, P., and Procaccia, I., 1983. Measuring the strangeness of strange attractors. Physica D, 9, 189-208.

Gupta, H.K. 1992. Reservoir-induced earthquakes. Elsevier, Amsterdam.

Gupta, H.K., Rastogi, B.K., Narain, H., 1972. Common features of reservoir associated seismic activities. Bull. Seism. Soc. Amer., 62, 481-492.

Heaton, T.H. 1975. Tidal triggering of earthquakes. Geophys. J. Roy. Astr. Soc., 43, 307-326.

Huang, L.Sh., McRaney, J., Teng, T.L., and Prebish, M. 1979. A preliminary study on the relationship between precipitation and large earthquakes in Southern California. Pure Appl. Geophys., 117, 1286-1300.

Kafri, V., and Shapira, A. 1990. A correlation between earthquake occurrence, rainfall and water level in Lake Kinnereth, Israel. Phys. Earth Planet. Int., 62, 277-283.

Lee, M., and Wolf, L. W. 1998. Analysis of fluid pressure propagation in heterogeneous rock-Implication for hydrologically induced earthquakes. Geophys. Res. Lett., 25 (13), 2329- 2332.

McClellan, P.H. 1984. Earthquake seasonality before the 1906 San Francisco earthquake. Nature, 307, 153-156.

McCloskey, J. 1993. A hierarchial model for earthquake generation on coupled segments of a transform fault. Geophys. J. Int., 115, 538-551.

McCloskey, J., Bean, C.J., and Jacob, A.W.B. 1991. Evidence for chaotic behavior in seismic wave scattering. Geophys. Res. Lett., 18, 1901-1904.

Mil'kis, M.R. 1986. Meteorological precursors of strong earthquakes. Phys. Solid Earth, 22(3), 195-204.

Muco, B. 1995. The seasonality of Albanian earthquakes and the cross- correlation with rainfall. Earth Planet. Int., 88, 285-291.

Muco, B. 1999. Statistical investigation on possible seasonality of seismic activity and rainfall-induced earthquakes in Balkan area. Phys. Earth Planet. Int., 114, 119-127.

Osborne, A.R., and Provenzale, A. 1989. Finite correlation dimension for stochastic system with power law spectra. Physica D, 35, 357-381.

Rai, S.S., Singh, Sunil K., Rajagopal Sarma, P. V. S.S., Srinagesh, D., Reddy, K.N.S., Prakasam, K.S., and Satyanarayana, Y. 1999. What triggers Koyna region earthquakes? Preliminary results from seismic tomography digital array. Proc. Indian Acad. Sci. (Earth Planet. Sci.), 108, 1, 1-14.

Schuster, A. 1897. On lunar and solar periodicities of earthquakes. Proc. R. Soc. London, 61, 455-465.

Simpson, D. W. 1986. Triggered earthquakes. Ann. Rev. Earth Planet., 14, 21-42.

Srivastava, H.N., Bhattacharya, S.N., and Sinha Roy, K. 1994. Strange attractor dimension as a new measure of seismotectonics of earthquakes around Koyna reservoir, India. Earth. Planet. Sci. Lett., 124, 57-62.

Stothers, R. B. 1989. Seasonal variations of volcanic eruption frequencies. Geophys. Res. Lett., 16, 453- 455.

Sugihara, G., and May, R.M. 1990. Non-linear forecasting as a way of distinguishing chaos from measurement error in time series. Nature, 344, 734-749.

Takens, F. 1981. Detecting strange attractor in turbulence. In: D. Rand and L.-S. Young (eds.), Dynamical Systems and Turbulence, Lecture Notes in Mathematics, pp. 366-381, Springer-Verlag, New York.

Talwani., P., and Acree, S. 1984. Pore pressure diffusion and the mechanism of reservoir-induced seismicity. Pure Appl. Geophys., 122 , 947-965.

Tiwari, R.K. and Rao, K.N.N. 1999. Phase space structure, Lyapunov exponent and non-linear prediction from earth's atmospheric angular momentum time series. Pure and Applied Geophysics, 16, 719-736.

Tiwari, R.K., and Rao, K.N.N. 2001. Power law random behaviour and seasonality bias of northeastern India earthquakes. J. Geol. Soc. India, 57, 369-376.

Tsonis, A.A., and Elsner, J.B. 1992. Non-linear prediction as a way of distinguishing chaos from random fractals. Nature, 358, 217-220.

8

Earthquake-Related Changes in Well Water Level and their Relation to a Static Deformation Model for the Seismically Active Koyna-Warna Region, India

Rajender Kumar Chadha, Kirti Srivastava and Hans-Joachim Kümpel

ABSTRACT

Coseismic water level rise related to a M 4.4 earthquake on April 25, 1997, was observed in two of the ten continuously monitored wells in the seismically active Koyna-Warna region in western India. The water level rise was in the form of step-like anomalies of size 7 and 3 cm at two wells located within 2 to 3 km of the epicenter. We calculated the coseismic strain field at the surface of a homogeneous elastic half space to understand the sign and amplitudes of the observed coseismic steps. The results show that while the sign of the water level steps is in agreement with the computed regional strain field, the observed amplitudes are either higher or lower than the ones predicted from the wells' volumetric strain responses. Anomalous precursory changes before this earthquake were not observed. From our observations we infer that coseismic strains in this area are a function of site effects controlled by local heterogeneities in the geological structures and simple that linear elastic models cannot explain the variations of the amplitudes of hydrological anomalies.

8.1 INTRODUCTION

In this chapter, we investigate changes in well water level with a M 4.4 earthquake in the Koyna-Warna region in western India. Several cases of anomalous water level changes in wells have been reported worldwide suggesting causal relationship between water level changes and earthquake occurrence in near and far fields (e.g., Wakita 1975; Igarashi & Wakita 1991; Koizumi et al., 1996; Ohno et al., 1997; Roeloffs & Quilty 1997; Roeloffs 1998; Grecksch et al., 1999). Roeloffs (1996) provided a comprehensive review on hydrological precursors and explained earthquake related hydrological phenomena in terms of poroelastic models. It has long been hypothesized that water level in wells connected to a confined aquifer, change in response to volumetric strain, with a coefficient of proportionality that can be determined by analysing water level fluctuations caused by earth tides. Though, understanding of the mechanism responsible for earthquake related hydrological anomalies is nowhere near to predictive levels, some basic ideas about the physical processes do exist. For example, coseismic changes in well water levels are believed to reflect sudden pore pressure changes, related to in-situ volumetric strain changes and the redistribution of stress in the brittle crust (e.g. Bodvarsson 1970; Kümpel 1992; Muir-Wood & King 1993; Koizumi and Tsukuda, 1999). Wakita (1975) explained the coseismic changes which followed the 1974 Izu-Hanto-Oki earthquake in Japan by a quadrant pattern of rising and falling of well water levels depending on the location of wells in either contractional or extensional regimes of induced volumetric strain. Later, Grecksch et al. (1999) also reported earthquake related water level changes to be consistent with the sign of calculated coseismic strain fields whereas in the case of Koizumi et al. (1996), the distribution of the signs of the coseismic groundwater level changes did not coincide with that of the volumetric strain changes.

Thirteen bore wells were drilled during 1995-1996, around the seismically active zone in the Koyna-Warna region. The depths of these bore wells range between 130 to 250 metres. By early 1997, 10 wells were instrumented with continuously operating pressure transducers of 1 mm resolution and digital recorders. Our studies also include systematic search for anomalous changes in water levels related to non-seismogenic crustal strains. This will however, not be discussed here. The seismic activity in the Koyna-Warna region started immediately after the initial impoundment of the Shivaji Sagar reservoir behind the Koyna Dam during early 1960s. The earthquakes continue to occur till date. The largest event that was associated with this reservoir was M 6.3 on December 10, 1967. Since then, the region has experienced more than 150 earthquakes of M ≥ 4.0. Several workers have attempted to explain the causative mechanism(s) for the seismic activity in the region (e.g., 8 Rajendran & Harish 2000; 1 Gupta 1992; 4 Gupta et al.,

1997; 3 Chadha et al., 1997; 7 Rai et al., 1999; 5 Rastogi et al., 1997; 2 Talwani et al., 1996, 6 Talwani 1997; Srinagesh et al., 2000).

We here report a coseismic step related to a local M 4.4 earthquake on April 25, 1997 as observed at two wells and compare it with the crustal volumetric strain computed for a homogeneous elastic half space. The earthquake occurred at 16:22 (UTC) at a hypocentral depth of about 5 km.

8.2 DATA ANALYSIS

Fig. 8.1 shows the location of the 10 wells where water level monitoring began during 1995 (Gupta et al., 2000). The well sites numbered 1 to 10 in Figure 8.1 are Koyna (KOY), Rasati (RAS), Kondavalle (KON), Taloshi (TAL), Vajegaon (VAJ), Govare (GOV), Chapher (CHA), Shringarpur (SHR),

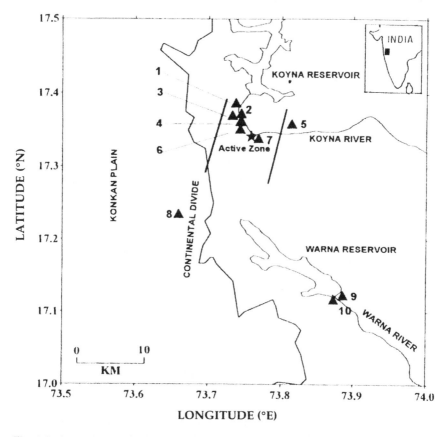

Fig. 8.1 Location of the Koyna-Warna reservoirs in western India. Solid triangles are the wells where water level fluctuations were monitored in 1997. A solar shows the epicenter of the M 4.4 earthquake on April 25, 1997.

Mandoor (MAN), and Ukalu (UKA). The wells were instrumented with pressure transducers sampled at 15 min interval. Other parameters monitored at few places only are the well water temperature, atmospheric pressure, air temperature, and rainfall. All the wells penetrate into the Deccan basalts, which cover the study area. The hydraulic transmissivity obtained for these wells from slug tests vary between 0.04 and 3.3 m^2/day.

Fig. 8.2a, b show water level data from all the 10 wells and atmospheric pressure, for the period March to May 1997. During the period of observation there was no significant rainfall. It is evident from the figures that KOY, GOV, TAL, VAJ, UKA, and MAN wells show clear spring and neap tidal signals indicating that they are connected to either confined or thick and lowly compressible aquifers. When analysing in more detail, we find that the response to earth tides in eight wells ranges from a minimum of 4 mm peak-to-peak in VAJ to 25 cm in UKA well. At CHA, SHR, and KON wells no tidal signals are observed in the data as can be quantified by regression techniques using a tidal catalogue and recordings of the barometric signal. To clean the water level recordings from variations in tidal strain and barometric pressure we have used the least squares regression scheme of Wenzel (1996). We analysed the data of all the 10 wells for the period 1995-1997. The annual rainfall in the region is up to 5,500 mm. Most of the precipitation occurs in the monsoon period during June-September, every year.

Figures 8a, b show water level data 15 days before and after the M 4.4 earthquake on April 25, 1997 after the removal of tides and atmospheric pressure effects. Clear coseismic steps of about 7 and 3 cm in water level height are seen at two wells, viz., TAL and GOV. No such steps are seen at other wells. The epicenter was located at 17° 20.62′ N and 73° 45.53′ E which lies 2.5 km off TAL and 1.8 km off GOV wells (Fig. 8.1). The focal mechanism solution for this earthquake was found to be left lateral strike slip on a plane striking NNE-SSW and dipping 55° NW as obtained from data of a closely operated seismic station network in the Koyna-Warna region (Rastogi et al., 1997). Uncertainty in the focal depth is about 1 km and is roughly <15 degrees for the orientation of the focal plane. The coseismic water level changes at TAL and GOV wells were measured as the maximum excursion in data corrected for tidal and barometric pressure variations. The anomalous water level heights were apparent in the first data sample following the earthquake origin time. No significant rainfall occurred prior or during this period. Also, these wells are in areas, which are not affected by human activities such as pumping of ground water. Unusual changes in the water level curves during the days prior to this earthquake were not observed. Also, no anomalous changes were noticed either in the well water temperature or air temperature in these regions.

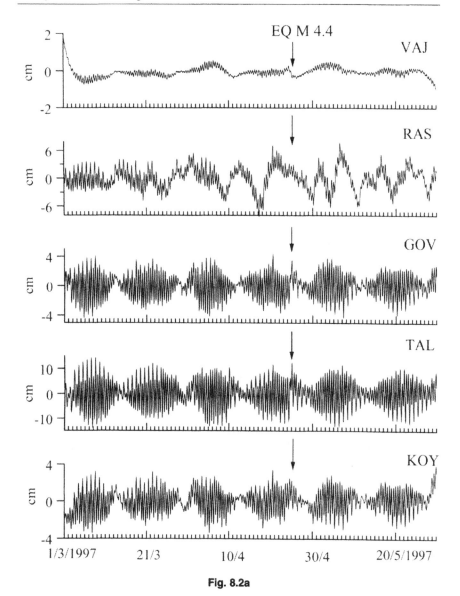

Fig. 8.2a

8.3 STATIC DEFORMATION

Well level fluctuations indicate pore pressure changes in the connected aquifers. Earth tides, variations in atmospheric pressure, fault creep, and passing of seismic waves are phenomena, which can deform an aquifer to cause changes in water levels (e.g., Wesson 1981; Van der Kamp & Gale, 1983; Rojstaczer, 1988; Liu et al., 1989).

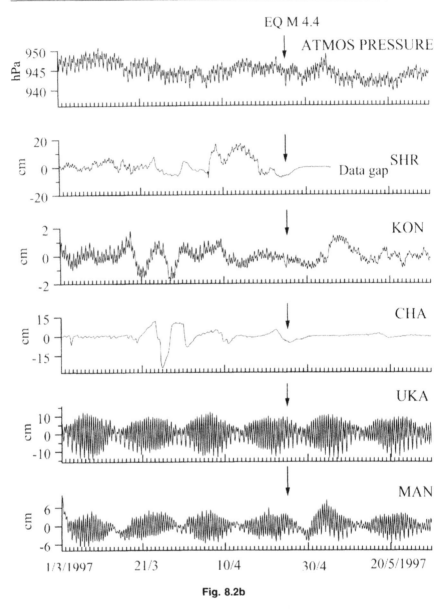

Fig. 8.2b

Fig. 8a, b Water level fluctuations recorded at 10 wells during March-May, 1997 in the Koyna-Warna region. Linear trends are removed from recordings. Atmospheric pressure recorded in the region is also shown. Spring and neap tides are clearly seen in the recordings of wells KOY, TAL, GOV, VAJ, UKA, and MAN and less significant in RAS. No tides were observed in KON, CHA, and SHR well's data.

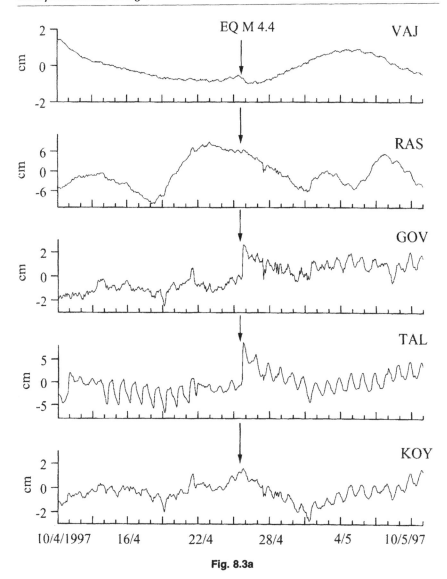

Fig. 8.3a

A coseismic change due to an earthquake thus overrides these changes and can produce a step-like change, which may only be seen if other effects are removed. According to a linear model that confirms the observations of Wakita (1975), the water level in the wells shall rise or fall based on the location of the wells with reference to the disposition of the fault plane. To see whether the signs of the coseismic steps observed at TAL and GOV due to the M 4.4 earthquake on April 25, 1997 agree with this model, we calculated

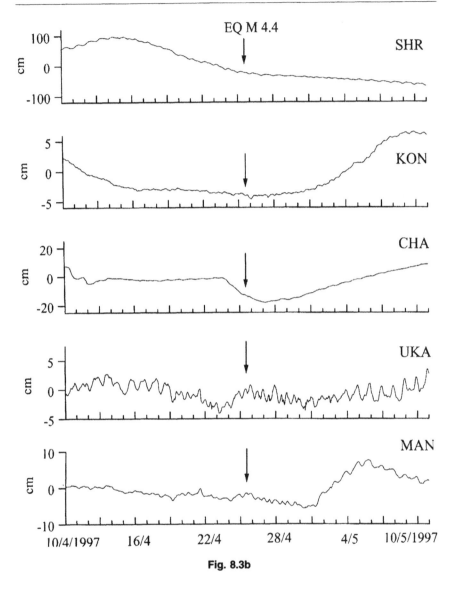

Fig. 8.3b

Fig. 8.3 a b   Water level fluctuations 15 days before and after the M 4.4 earthquake on April 25, 1997, after the removal of tides and barometric pressure. Coseismic steps are seen at Taloshi (7 cm) and Govare (3 cm).

the static volumetric strain at the surface of a half-space model. To quantify the static deformation field in a given region, Okada (1992) has given complete sets of closed form analytical expressions that have been derived for the internal deformation, surface deformation and strains for a point and a rectangular source in different faulting environments. Using input source parameters such as source type, seismic moment, dip of the fault, P- and S-wave velocities, density of the medium, depth of the focus, and fault plane slip one can compute the displacements, strain, stress, and static volumetric strain. Here, we used parameters for this earthquake from different studies (Rastogi et al., 1997; Talwani 1997) and adopted some average density value and seismic P- and S-wave velocities to compute static volume strain at the surface using the Okada (1992) formulation (Table 8.1).

Table 8.1 *Parameters used for calculating the static coseismic strain field for a homogeneous elastic half-space model (after Okada, 1992).*

Focal mechanism	left-lateral strike-slip
Dip	55° NW
Strike	N 30° E
Seismic moment	1.0×10^{15} Nm
Focal depth	5 km
Displacement	5 cm
P-wave velocity, V_p	6,300 m/s
S-wave velocity, V_s	3,705 m/s
Density	2,760 kg/m^3
V_p/V_s	1.71

Fig. 8.4 shows the volumetric strain field computed on a 1 km × 1 km surface grid for a point source model of the April 25, 1997 earthquake. The expanding and contracting zones representing the hypothetic static volume strain field following this earthquake are clearly seen with negative values indicating compressional strain regime where water level should rise. TAL and GOV wells where coseismic steps exhibit water level rise are located in the compressional strain zone and hence, agree with the model. Table 8.2 shows the volumetric strain calculated for different focal depths for all the 10 wells. The strains are found to critically depend on the source geometry and the focal depth. Whereas, change of 1 km in focal depth shows a rather large variation in volumetric strain amplitude, a change in the dip angle of the fault by ± 10° has a less significant impact. The vertical displacements at the well locations were also computed and were found to show a similar pattern.

8.4 CONFINED VOLUMETRIC STRAIN EFFICIENCIES

The size of the expected steps in water levels, at a given well location can be estimated using linear theory of poroealsticity, assuming that affected aquifers behave elastically. Kümpel (1991) gave a comprehensive review

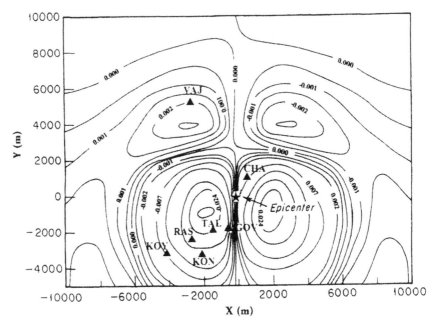

Fig. 8.4 Contours of the surface static volumetric strain field (μ strain) due to the April 25, 1997 earthquake calculated for a point source using the Okada (1992) formulation. Solid triangles show locations of wells. The epicenter is shown with a star. UKA, MAN, and SHR wells are not shown. X direction is parallel to the strike of the fault in the NNE-SSW direction.

Table 8.2 *Calculated volume strains (Δ_V) at all the wells for different focal depths*

No.	Well	Distance (km) from epicenter	Calculated Volume Strain (μ strain)		
			4 km	5 km	6 km
1.	KOY	5.11	− 0.002	− 0.007	− 0.008
2.	RAS	3.40	− 0.024	− 0.020	− 0.014
3.	KON	3.95	− 0.013	− 0.013	− 0.010
4.	TAL	2.50	− 0.041	− 0.024	− 0.014
5.	VAJ	6.00	0.001	0.002	0.001
6.	GOV	1.85	− 0.024	− 0.012	− 0.006
7.	CHA	1.30	0.006	0.004	0.002
8.	SHR	16.15	− 0.001	− 0.001	− 0.001
9.	MAN	28.04	0.000	0.000	0.000
10.	UKA	28.10	0.000	0.000	0.000

about the poroelastic parameters and Wang (1993) discussed various geophysical applications. Frequently used parameters that describe the deformation of the isotropic poroelastic media are the rigidity or shear modulus μ, the Skempton parameter B, and the Poisson ratios for drained and undrained conditions, v and n_u. Roeloffs & Quilty (1997) gave an expression for the strain-induced change in pore pressure under undrained conditions as

$$P = -B \cdot K_u \cdot \Delta_v \tag{1}$$

where the change in pore pressure P is described as being proportional to changes in volumetric strain, Δ_v. K_u is the undrained bulk modulus of the aquifer, which is given by $\dfrac{2\mu(1 + v_u)}{3(1 - 2v_u)}$ and the quantity BK_u is termed as the static confined volumetric strain sensitivity (Rojstaczer, 1988). The latter can be obtained from the analysis of tidally induced water level fluctuations in wells, since the tidal forcing is well known (for example, Wilhelm et al., 1997).

Roeloffs (1996) has given the change in water level Δh for slow changes in pore pressure in a confined aquifer as

$$\Delta h = -\Delta_v \frac{BK_u}{\rho_f g} \tag{2}$$

Herein, ρ_f is the fluid density, g is the gravitational acceleration. $\dfrac{\Delta h}{\Delta_v}$ is the static confined volume strain efficiency which is seen to be highly dependent on the rigidity modulus, the Skempton parameter and the undrained Poisson ratio. Uncertainties in these parameters can give rise to large uncertainties in the calculated values. The range of the Skempton parameter B is known to vary between 0 and 1. For water-saturated soils B is close to 1. The Poisson ratios scale as $0 \le v \le v_u \le 0.5$. All the parameters that hold for undrained condition are not routinely measured and hence are not available for the region.

Based on a deformable elastic earth model, strain sensitivities for all the 10 wells were obtained using ETERNA program for tidal analysis (Wenzel 1996). The values are shown in Table 8.3. Most useful tidal constituents for such analysis are O_1 and M_2 (diurnal and semi diurnal) because they yield highest signal-to-noise ratios and are not too biased by meteorological influences. Janssen (1998) obtained tidal strain sensitivities for seven well aquifer systems in the Koyna-Warna region from such tidal analysis. He showed that signal-to-noise ratios are higher than 15 for M_2 amplitudes for wells GOV, KOY, MAN, RAS, TAL, UKA, and VAJ. Minimum and maximum lengths of the recordings taken for his analysis are 101 and 183 days, respectively. Using these strain sensitivities we applied equation (2) to estimate the amplitudes of the expected coseismic steps at each well (Table 8.3). In view of the sensitivity values for each well, and taking into account

Table 8.3 *Expected and observed well water level response at all the 10 well locations. Strain sensitivities at each well are calculated from the tidal analysis ETERNA 3.3 software (Wenzel, 1996). Expected change in well water level response is calculated for 5 km which is focal depth of the earthquake. Nominal resolution of pressure transducer for well level recording is 1 mm.*

No.	Well	Strain sensitivity from tidal analysis (mm/nano strain)	Observed change in well water level (cm)	Expected change in well water level (cm)
1.	KOY	−0.71 ± 0.02	not observed	0.48 ± 0.016
2.	RAS	−0.40 ± 0.08	not observed	0.81 ± 0.15
3.	KON	low signal-to-noise ratio	not observed	—
4.	TAL	−3.9 ± 0.20	7.0	9.5 ± 0.49
5.	VAJ	−0.06 ± 0.003	not observed	0.01 ± 0.0015
6.	GOV	−0.93 ± 0.02	3.0	1.09 ± 0.03
7.	CHA	low signal-to-noise ratio	not observed	—
8.	SHR	low signal-to-noise ratio	not observed	—
9.	MAN	−1.66 ± 0.10	not observed	0.02 ± 0.002
10.	UKA	−4.15 ± 0.08	not observed	0.06 ± 0.003

the epicentral distances of the wells, strongest well level changes in response to the M 4.4 earthquake on April 25, 1997 could be expected for TAL, GOV, KOY, and RAS wells. Wells at MAN and UKA are nearly 5 times more distant and CHA, KON, and SHR wells show absence or very low signal-to-noise ratios for tides, hence coseismic steps were not expected to be observed at these wells.

From Fig. 8.3 a & b it is seen that coseismic steps of 7 cm and 3 cm were observed at TAL and GOV wells. Although there was a small water level rise in KOY and RAS wells after the earthquake, no step-like anomaly was observed at these wells. Table 8.3 displays the discrepancy between the observed and the expected amplitudes at TAL and GOV wells. While at TAL the observed step was found to be 2.5 cm lower than the calculated one, it was higher by 2 cm in case of GOV well. It is also seen that unlike TAL well, GOV is located close to a strong gradient in static volumetric strain (Fig. 8.4). A slight change in the strike angle of the focal plane of the forcing earthquake could result in strongly different amplitude of well level response and thus explain the discrepancy here. Also, both TAL and GOV wells may respond to some additional process rather than the elastic strain. At VAJ, MAN, and UKA wells no noticeable well level anomalies were observed which agrees with the strain sensitivities of these wells and the elastic model. Moreover, predicted amplitudes are less than 1 mm which are below the resolution of the transducers.

8.5 DISCUSSION AND CONCLUSION

The M 4.4 earthquake on April 25, 1997, has provided a set of good observations to test the hypothesis that "well level fluctuations respond to changes in crustal volume strain that is induced by an earthquake in the form of step like coseismic changes". Out of the 10 wells that were monitored coseismic step was observed in only two wells at TAL and GOV. Other wells did not show any step-like response to this earthquake. Although several earthquakes of magnitude less than 4 occurred during the study period 1995-1997 no noticeable coseismic step was observed. To see whether the coseismic steps observed at the two wells agree with the observations of Wakita (1975) that the water level in a well shall rise or fall based on the location of the well with reference to the disposition of the fault plane, we used Okada's (1992) solutions to calculate the static volumetric strain at the surface of a homogeneous half-space model. Our results show expanding and contracting zones representing the static volumetric strain field following this earthquake. TAL and GOV wells where coseismic step-like rise was observed are located in the compressive zone and hence, in the sign agree with the model. From this it is inferred that well aquifer systems indeed have the potential to reflect coseismic volumetric strain changes, similar as with tidal fluctuations in volume strain. As most of our wells show good tidal response, we calculated strain sensitivities for each well using the tidal analysis scheme of Wenzel (1996). Using these strain sensitivities we estimated the amplitudes of the expected well level changes at each well. Our results show that the misfit in amplitudes of the observed and the calculated coseismic steps does not show any pattern. While the observed coseismic step is 2.5 cm less than the calculated in case of TAL it is 2 cm more in GOV well. In other wells where the predicted amplitudes are less than 1 mm, we did not observe any noticeable well level anomalies and hence, agree with the elastic model. From these observations we infer that the coseismic strains may be a function of site effects controlled by local heterogeneity in geological structures. It is probably not surprising that simple elastic models cannot explain the amplitudes of hydrological anomalies, wholly. Our results, thus, agree with the findings of earlier workers (e.g., Roeloffs 1988; Koizumi et al., 1996; Wang 1997). Huang et al. (1995) suggested a non-linear response of water levels to coseismic strains due to local heterogeneities. Grecksch et al. (1999) observed coseismic well level steps to be higher than predicted from strain efficiencies.

Continuous data from a close network of wells encompassing a seismogenic zone in the Koyna-Warna region has not shown any hydrological anomalies for earthquakes below the M < 4.0 level so far. Our present studies indicate that co- and post-seismic changes in the well water levels can be expected for moderate earthquakes in the region. Over the years well level recordings from the seismically active Koyna-Warna region may provide valuable data sets for doing more quantitative analysis to understand the processes involved in producing hydrological anomalies in general and the role of pore fluid in crustal rheology, in particular.

ACKNOWLEDGEMENTS

The authors wish to thank Dr. H.K. Gupta, Secretary, Department of Ocean Development, New Delhi for his encouragement and valuable suggestions for this work. We thank Prof. Y. Okada for providing us his program routines. Support from the Council of Scientific and Industrial Research, New Delhi, India and the Federal Ministry of Education and Research of Germany through DLR/International Bureau, Bonn for this work is gratefully acknowledged.

REFERENCES

Bodvarsson, G. 1970. Confined fluids as strain meters. J. Geophys. Res: 75, 2711-2718.

Chadha, R.K., Gupta, H.K., Kümpel, H.-J., Mandal, P., Nageswar Rao, A., Radhakrishna, I., Rastogi, B.K., Raju, I.P., Sarma, C.S.P., Satyamurthy, C., and Satyanaryana H.V.S. 1997. Delineation of active faults, nucleation process and pore pressure measurements at Koyna (India). Pure Appl. Geophys: 150, 551-562.

Grecksch, G., Roth, F., and Kümpel H.-J. 1999. Co-seismic well level changes due to the 1992 Roermond earthquake compared to static deformation of half space solutions. Geophy. J. Int, 138, 470-478.

Gupta, H.K. 1992. Reservoir-Induced Earthquakes., Elsevier Scientific Publishing Co., Amsterdam, p 355.

Gupta, H.K., Rastogi, B.K., Chadha, R.K., Mandal, P., and Sarma, C.S.P. 1997. Enhanced reservoir-induced earthquakes in Koyna region, India, during 1993-95, J. Seismol, 1, 47-53.

Gupta, H.K, Radhakrishna, I., Chadha, R.K., Kümpel, H.-J., and Grecksch, G. 2000. Pore pressure studies initiated in area of reservoir-induced earthquakes in India. EOS, Trans. Am. Geophys. Un 81, 81(14) 145 and 151.

Huang, W., Rojstaczer, S., and Breau, S. 1995. Coseismic response of water level to earthquakes in the San Jacinto Fault, Southern California. EOS, Trans. Am. Geophys. Un., 76 (Suppl.), 359.

Igarashi, G. and Wakita, H. 1991. Tidal responses and earthquake-related changes in the water level of deep wells. J. Geophys. Res, 96, 4269-4278.

Janssen, J. 1998. Untersuchung des Zusammenhangs zwischen lokaler Seismizität und dem Porendruck in gespannten Aquiferen in der Koyna-Region, Indien. Dipl.thesis, Institute of Physics, Univ. of Bonn, Germany, pp120.

Koizumi, N., Kano, Y., Kitagawa, Y., Sato, T., Takahashi, M., Nishimura, S., and Nishida, R. 1996. Groundwater anomalies associated with the 1995 Hyogo-ken Nanbu earthquake. J. Phys. Earth, 44, 373-380.

Koizumi, N. and Tsukuda, E. 1999. Pressure changes in groundwater level and volumetric strain associated with earthquake swarms off the east coast of the Izu peninsula, Japan Geoph. Res. Lett., 26(23), 3509-3512.

Kümpel, H.-J. 1991. Poroelasticity: parameters reviewed. Geophys. J. Int, 105, 783-799.

Kümpel, H.-J. 1992. About the potential of wells to reflect stress variations within inhomogeneous crust. Tectonophysics, 211, 317-336.

Liu, L.-B., Roeloffs, E. and Zheng, X.-Y 1989. Seismically induced water level fluctuations in the Wali Well, Beijing, China. J. Geophys. Res., 94, 9453-9462.

Muir-Wood, R. and King, G.C.P. 1993. Hydrological signatures of earthquake strain. J. Geophys. Res, 98, 22035-22068.

Ohno, M., Wakita, H., and Kanjo, K. 1997. A water well sensitive to seismic waves. Geophys. Res. Lett, 24, 691-694.

Okada, Y. 1992. Internal deformation due to shear and tensile faults in a half space. Bull. Seism. Soc. Am, 82(2), 1018-1040.

Rai, S.S., Singh. Sunil K., Rajagopal Sarma, P.V.S.S., Srinagesh, D., Reddy, K.N.S., Prakasam, K.S., and Satyanarayana, Y. 1999. What triggers Koyna region earthquakes? Preliminary results from seismic tomography digital array. Proc. Indian. Acad. Sci, 108(1), 1-14.

Rajendran, K and Harish, C.M. 2000. Mechanism of triggered seismicity at Koyna: An evaluation based on relocated earthquakes. Curr. Sci, 79 (3), 358-363.

Rastogi, B.K., Chadha, R.K., Sarma, C.S.P., Mandal, P., Satyanarayana, H.V.S., Raju, I.P., Kumar Narendar, Satyamurthy, C., and Rao, A. Nageshwar 1997. Seismicity at Warna Reservoir (near Koyna) through 1995. Bull. Seism. Soc. Am: 87(6), 1484-1494.

Roeloffs, E.A. 1988. Hydrologic precursors to earthquakes: a review. Pure Appl. Geophys, 126, 177-209.

Roeloffs, E.A. 1996. Poroelastic techniques in the study of earthquake related hydrological phenomena. Adv. Geophys, 37, 135-195.

Roeloffs, E.A. 1998. Persistent water level changes in a well near Parkfield, California due to local and distant earthquakes. J. Geophys. Res, 103, 869-889.

Roeloffs, E.A. and Quilty, E. 1997. Water level and strain changes preceding and following the August, 1985 Kettleman Hill, California, earthquake. Pure Appl. Geophys, 149, 21-60.

Rojstaczer, S. 1988. Determination of fluid flow properties from response of water levels in wells to atmospheric loading. Water Res. Res, 24, 1927-1938.

Srinagesh, D., Singh, S.K., Reddy, K.N.S., Prakasam, K.S., and, Rai S.S. 2000. Evidence for high velocity in Koyna Seismic zone from P wave Teleseismic imaging. Geophys. Res. Lett, 27(17), 2737-2740.

Talwani, P. 1997. Seismotectonics of the Koyna-Warna Area, India. Pure Appl. Geophy, 150, 511-550.

Talwani, P., Kumar Swamy, S.V., and Sawalwade, C.B. 1996. The revaluation of seismicity data in the Koyna-Warna Area, 1963-1995; Univ. of South Carolina, Tech. Rep, pp343.

Van der Kamp, G. and Gale, J.E. 1983. Theory of earth tide and barometric effects in porous formations with compressible grains. Water Res. Res, 19, 538-544.

Wakita, H. 1975. Water wells as possible indicators of tectonic strains. Science: 189, 553-555.

Wang, H.F. 1993. Quasi-static poroelastic parameters in rock and their geophysical applications. Pure Appl. Geophys: 141, 269-286.

Wang, H.F. 1997. Effects of deviatoric stress on undrained pore pressure response to fault slip. J. Geophys. Res, 102, 17 943-17 950.

Wenzel, H.-G. 1996. The nanogal software: Earth tide data processing package ETERNA 3.30. Bull 'd Informations Marees Terrestres. 124, 9425-9439.

Wesson, R.L. 1981. Interpretation of changes in water level accompanying fault creep and implications for earthquake prediction. J. Geophys. Res, 86, 9259-9267.

Wilhelm, H., Zürn,W., and Wenzel, H.-G. 1997 (eds.). Tidal Phenomena, Lecture Notes in Earth Sciences, Vol. 66, Springer, Berlin.

9

Geothermal Energy—An Energy Option for the Future

Fritz Rummel

ABSTRACT

Geothermal energy presently is used in the form of direct heat for various purposes, and to generate electricity. Direct use of heat can be estimated as 20,000 MW (thermal), the installed geothermal electricity capacity is almost 10,000 MW (electric), worldwide. For some countries with favourable geological conditions geothermics could become the major energy source in the near future. Since heat is stored everywhere within the upper continental crust in large quantities, geothermal energy may also provide an option to other countries when we consider the potential and develop tools for exploitation.

9.1 INTRODUCTION

Progress in civilization requires the disposability of energy. For our ancestors, fire was protection against wild animals and for survival in cold regions. Much later heat was converted into mechanical and electrical energy and became the basis for the development of our modern science and technology society. Presently, the annual world energy consumption can be estimated as about 4×10^{20} Joule which corresponds to about 13 billion tons of coal or 9 billion tons of oil equivalent (1 J $\equiv 3.41 \times 10^{-11}$ tCE $\equiv 2.35 \times 10^{-11}$ tOE). Approximately one-third of this energy is consumed by

6 per cent of the world population in highly industrialized countries. The per capita annual electricity consumption is less than 200 kWh in India and about 7,000 kWh in Germany or 11,000 kWh in the USA. Inspite of intensive efforts for energy savings in industrialized countries particularly since the Kyoto World Climate Conference in 1997, the world energy demand will rapidly increase in the near future due to population explosion, progressive urbanization, industrialization, and people's request for better living conditions in developing countries. In addition, millions of people in countries like China expect fuel for vehicles to make use of their recently gained mobility, electric power is necessary for the increasing production demand of raw materials, and also for communication systems.

Until present our energy supply is mainly based on the world coal, oil, and gas resources. These are finite and will last for some decades. However, these materials are much too valuable to be burnt for energy conversion. The present ongoing process of political and economical liberalization will lead to conflicts with respect to the access and the distribution of fossil fuels and will result in reformations on the energy market at dimensions never experienced in the past. Nuclear power has gained some importance but is associated with possible reactor hazards, the misuse of radioactive components, and the unsolved problem of safe long-term waste disposal. The further development of water power is nearly exhausted worldwide. Renewable energies such as wind, tidal, and biomass demonstrate lacks related to local and temporal availability, solar energy is still in the early development, and nuclear fusion still requires heavy research and financial investments. Thus, what are the mid-term and long-term energy options (Fig. 9.1). The question is a complex challenge which is addressed by various prognosis and energy scenarios which consider "new horizons" with positive liberalization effects or "barriers" with limited liberalization due to regional, economic, national, cultural, or religious isolations (Shell, 2000). The solution is far beyond the capacity and responsibility of a single nation, and the development of a new energy technology for the market requires 50 to 100 years which was valid in the past for coal, oil, gas, and nuclear energy (Fig. 9.1). The question arises whether geothermal heat could play an important role in such energy scenarios.

9.2 GEOTHERMAL ENERGY POTENTIAL

The temperature within the earth's interior can be estimated to range between 3000 and 5000 centigrades. The large uncertainty simply originates from the fact that temperatures cannot be measured directly within the earth. The indirect estimates rely on the melting temperatures

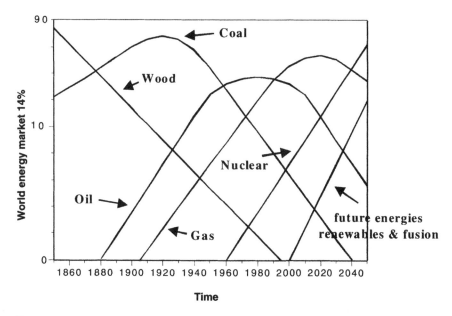

Fig. 9.1 The Marchetti diagram illustrates the substitution processes of primary energy supply

for silicates in the solid mantle and for nickel/iron in the molten earth's core at high pressure, or on assumptions of the adiabatic temperature gradient at depth. In comparison, the temperature distribution within the earth's crust can be given more precisely by extrapolating the thermal gradient of approximately 30°C per kilometre measured in superdeep boreholes to a depth of about 10 kilometres. Thus, we can expect a temperature of 200°C at a depth of 7 km for most continental regions. Geothermal anomalies with much higher geothermal gradients, often associated with volcanic activity generally are located along lithospheric plate boundaries. In Iceland, e.g. on the mid-Atlantic ridge or in Japan or New Zealand on the Circum-Pacific plate boundary we find subsurface temperatures of some hundred centigrades at shallow depth. On the other hand old continental crust regimes like the Indian subcontinent, central Eurasia or the central African crust are charcterized by low thermal gradients. The rock temperature in the deep gold mines in southern India or South Africa is merely 60°C at a depth of 3 km.

Speculating about the origin of the earth, its thermal history, and its present heat content we may consider the following energy terms:

- The gravitational energy originates from the rapid contraction of gas and dust particles during the very early stage of the formation of our planet Earth about 4.5 billion years ago. With all uncertainties about this mass accretion it may have occurred during a short period of 200 million years. It may be assumed that most of this energy was converted into heat. The energy influx at the end of this accretion phase can be expressed by a term of the form

$$\frac{dE_i}{dt} = \frac{f\bar{\rho}}{3} \times \frac{\Delta m_i}{\Delta_t} \tag{1}$$

where f is the gravitational constant, $\bar{\rho}$ is the mean density of the accumulated earth mass, and $\Delta m_i / \Delta_t$ is the rate of mass increase by meteorite impacts at that time.

- The heat production of short-life radioactive isotopes, particularly Al^{26} and Fe^{60} may have had some importance for a few million years during this initial period, similar to the heat exchange on the surface with respect to meteorite impacts, the heat consumption for melting, heat conduction and convection from the interior to the surface, and solar radiation to the earth.
- Heat radiation into space can be described by the Stefan-Boltzmann law

$$\frac{dE_i}{dt} = \sigma \times T^4 \tag{2}$$

where σ is the Stefan-Boltzmann constant and T is the absolute temperature.

By neglecting most terms we may then make an energy balance between the heat content of the primitive earth and the heat losses due to radiation which yields the equation:

$$\frac{f\rho}{3} \frac{\Delta m}{\Delta t} = \sigma T^4 \tag{3}$$

This allows us to speculate the mean temperature of the earth at the end of its formation. Using relevant numbers we can conclude that the Earth was quite cool and probably mostly solid. We may then estimate the heat content of the earth between 15 and 35×10^{30} Joule, the low value for a cold solid earth corresponding to a slow rate of mass accumulation, the high value for a hot virgin earth accumulated by fast rate mass accretion. Subsequently to this stage, radiation balance between solar influx and radiation from the earth could begin. Rather rapidly a mean surface temperature of about 14°C was reached which remained essentially constant until today.

The crustal temperature profile is determined by the heat production of natural radioactive isotopes concentrated in the earth's crust on the one hand, and the heat loss at the earth's surface on the other hand. The heat production is mainly due to the decay of the uranium and thorium isotopes U^{235}, U^{238}, Th^{232}, and of the potassium isotope K^{40}. Neglecting the uncertainty on the concentration and distribution of these isotopes we may estimate that in the granitic upper crust radiogenic heat is produced at a rate of 3 microWatts per cubicmetre. This heat approximately compensates the heat loss at the earth's surface at a mean heat flow density rate of 55 ± 5 mW/m^2 as measured for continents. This terrestrial heat flow at the Earth's surface is more than 20,000 times smaller than the solar influx of 1.35 kW/m^2. The excess heat flux from the sun is the source for photosynthesis (biomass) and other energies (wind, sea tides).

The small terrestrial heat flow which radiates an energy of 10^{21} Joule per year into the atmosphere, however, is too small to be directly used as energy source. On the other hand, the stored heat within the upper 10 km of the continental earth's crust is about 6×10^{26} Joule if we assume a mean temperature of 150 centigrades, a specific heat of crustal rock of about 800 Joule per kilogram and per centigrade, and a mean rock density of 2,700 kg/m^3. This energy would be sufficient to operate a million power plants of a capacity of 200 MW for 10,000 years. A more realistic example: a cubic kilometre of hot granite at a depth of about 5 km could provide a thermal energy of 2×10^{17} Joule if cooled by 100°C, to operate a 30 MW electric power plant for 30 years.

9.3 GEOTHERMAL SYSTEMS

Geothermal anomalies and phenomena like hot springs or volcanism are evidences for the presence of geothermal energy within the earth's crust. There origin requires the simultaneous existence of very specific geological boundary conditions which generally only are given at lithospheric plate boundaries in ridge, trench, or graben systems or in continental collision zones. The geological conditions are characterized by deep fracture systems which allow heat ascent by fluid convection, porous rock formations for storage of thermal fluids, and an impermeable rock cover for thermal and hydraulic isolation. However, as mentioned above, the hot crystalline rocks at greater depth per se present an enormous heat source. In practise the following geothermal systems for energy use may be differentiated (Table 9.1):

- **Warm water** may exist in shallow, highly permeable and porous/ fractured sedimentary formations at a depth of some tens or hundreds of metres with temperatures up to say 60°C. It may be directly used for space heating, as process heat, in green houses, or

Table 9.1   *Status of geothermal energy use*

Type	Technological Level	Problems
warm water (<90°C) from shallow systems	developed	high investment for small consumers, use only for heat
hydrothermal hot water/steam systems (90-250°C)	fully developed for electricity production	limited resources at few locations, salt content, corrosion
geopressurized systems in deep sediment basins	exploration	off-shore, methane content, subsidence
volcanic systems	pre-exploration	new technologies required
hot-dry-rock systems - at location with high thermal gradients	research and development stage for electricity or combined use	stimulation techno-logy, induced seis-micity, circulation fluid losses, only
-at locations with normal thermal gradients	first research for few projects big consumers	mostly for heat only, drilling costs, frac technology, requires

for balneology. Generally, such reservoirs are used by closed-loop doublet-operations where heat is extracted from the thermal water by a heat exchanger at surface and the cold water is reinjected into the aquifer. Although no exact figures exist the worldwide installed thermal capacity of warm water above a temperature level of 35°C can be estimated to about 10,000 MW (thermal). In Europe, the most prominent example of warm water use for heating is in the Parisian basin where approximately 200,000 homes are heated. Another example is in the North German basin where several geothermal heat plants for space heating with a total capacity of about 20 MW (thermal) were installed since 1984. As an example for the use of thermal water for balneology, the thermo-therapeutic recreation area around Bad Füssing in south-eastern Bavaria/Germany may be mentioned. Besides geothermal aspects balneology may include social-economical effects which should not be underestimated. In the Bad Füssing area more than one million visitors per year swim and relax in thermal water. In 1994 the worldwide installed capacity for direct use of warm water systems for space heating, process heat, snow and ice melting, air conditioning, balneology, and in agriculture was estimated as 8,600 MW (th) (Table 9.2) or as an annual thermal energy

Table 9.2 *Direct use of low enthalpy geothermal energy in 1995 (after Freeston, 1995)*

Country	Thermal Power MW (th)	Major Use
Austria	21.1	space heating, balneology
Belgium	3.9	swimming, fish forming
Bulgaria	133	space heating, greenhouses
China	2,143	various purposes
Croatia	15	heating, balneology
Denmark	3.5	
France	337	space heating
Georgia	245	heating, green houses, balneology
Germany	60 (app.)	space heating, balneology
Greece	23.2	green houses
Hungary	340	balneology, agriculture
Iceland	1,443	space heating, process heat
Israel	44.2	agriculture
Italy	308	balneology, heating
Japan	318	space heating, greenhouses
Netherlands	2	heat pumps
New Zealand	264	process heat
Poland	63	air conditioning
Romania	137	heating, bathing, greenhouses
Russia	210	green houses, space heating
Serbia	80	not known
Slovakia	100	not known
Slovenia	37	swimming
Sweden	47	heat pumps
Switzerland	110	heat pumps
Turkey	140	space heating, balneology, greenhouses
USA	1874	heat pumps

Total 8500 (app.)

consumption of 10^{17} Joule (Freeston, 1995). The numbers are derived from data of 23 countries and, therefore, are an underestimate. Clearly, some countries stand out as major consumers for direct use such as China, the USA, Japan, and Iceland where 85 per cent of the houses are heated with geothermal water. However, it must also be recognized that such geothermal developments have proceeded at only a slow pace during the past 10 years since low oil and gas prices provided cheaper options to developers in most countries.

- **Hot water** or **steam dominated hydrothermal convective geothermal aquifers** with temperatures up to 250°C at depths between 1 and 3 km originate by the formation of deep sedimentary rocks with high pore water content and by heating the pore water from a deep-seated heat source. For thermal/hydraulic isolation the aquifer must be sealed by an impermeable sedimentary cover on its top.

At present all of the existing geothermal power plants are using heat from such hydrothermal high-enthalpy system located on plate boundaries (Fig. 9.2). The total installed capacity was 6,800 MW (e) (in 1995) and may have increased to approximately 10,000 MW (e) in the year 2000 (Table 9.3). The development started in 1912 at Lardorello in the Italian Tuscany where the Prince Piero Ginori Conti installed the first 250 kiloWatt electric generator. Today, Italy produces app. 632 MW (e) from the Tuscany hydrothermal system. The most prominent example for geothermal power production is in the Californian geothermal field "The Geysers", which supplies energy to the cities of San Francisco and Oakland. Recently, some parts of the Geyser field show evidences of gradual exhaustion and require deepening and stimulation. The fastest development occurred in the Philippines from five geothermal fields on Luzon and in the Visayas Islands. The installed capacity was 890 MW (e) in 1985, 1,227 MW (e) in 1995, and may have reached 2,000 MW (e) today. The power plant sizes range from 11 MW (Tiwi plant) to 345 (Mak-Ban). Rapid developments can be expected for the central American countries like Mexico, El Salvador, Nicaragua, and Cost Rica where geothermal energy may become the major energy source. In comparison, power production in Japan was only 300 MW (e) in 1994 with a goal to achieve a total capacity of 600 MW (e) in the year 2000. Although it is recognized that geothermal is an important national energy source, the present use is only 0.4 per cent of the total power production in Japan (Yamaguchi et al., 1995). Even the direct use of geothermal heat for space heating, process heat or balneology was surprisingly small, app. 318 MW (th) in 1994 (Sekioka and Toya, 1995), inspite of favourable geothermal conditions.

- In very young crustal regimes with extension tectonics we find **volcanic geothermal systems** with hot water and steam reservoirs in the subsurface and magma chambers at shallow depth. Estimates show that in the upper 7 km of the continental crust an energy of 5×10^7 megaWatt years is stored in such systems at temperatures of 1000°C. Tapping such magma reservoirs or extracting heat from boiling lava lakes on surface could be the most spectacular form of geothermal energy use. Various proposals have been made in the past to use this energy such as to

Fig. 9.2 Worldwide installed geothermal power plants along plate boundaries (numbers give installed power for electricity production in megawatts)

Table 9.3 *Installed geothermal power generation capacity in MW (e), (from : Huttrer, 1995)*

Country	1990	1995	2000[1]
Argentina	0.67	0.67	[2]
Australia	0	0.17	[2]
China	19.2	28.78	81
Costa Rica	0	55	170
El Salvador	95	105	165
France	4.2	4.2	[2]
Iceland	44.6	49.4	[2]
Indonesia	144.8	309.8	1080
Italy	545	631.7	856
Japan	214.6	413.7	600
Kenya	45	45	[2]
Mexico	700	753	960
New Zealand	283.2	286	440
Nicaragua	35	35	[2]
Philippines	891	1227	1978
Portugal/Azores	3	5	[2]
Russia	11	11	110
Thailand	0.3	0.3	[2]
Turkey	20.6	20.6	125
USA	2774.6	2816.7	3395
Total	**5832**	**6798**	**9960**

[1] estimated
[2] information not available

inject water and recover the produced steam or to insert special heat exchangers into the molten rock. Although these ideas are far from any present technical feasible realization, the energy source is attractive and obviously available at many locations in the vicinity of active volcanism, and requires totally new concepts.

- In highly porous rock formations which rapidly subsided to great depth together with the enclosed pore fluids and gas content (mostly methane) we find **geopressurized water systems** with high pore pressure and high temperatures. Such systems were first identified in the deep sedimentary layers underneath the Gulf of Mexico at depth between 6 and 8 km with pore pressures of 130 MPa and temperatures between 150 and 180°C. Some production tests from drillholes indicate a possible productivity of some millions of cubicmetres of hot brine per day from such systems, with a significant amount of hydrocarbon gas. The energy content of this specific system consisting of hot pressurized fluids and the hydrocarbons is estimated as 5×10^9 megaWatt years. Similar systems may exist in other deep sedimentary basins like in the Po basin in Northern Italy, the Molasse basin in southern Germany, or in Northern India.

- By far, the largest amount of heat is stored in the hot crystalline basement of the upper continental crust. As mentioned before, at a depth of 6 km we can expect a temperature of 200°C almost everywhere. High pressure of say 160 mega-Pascals at such depth prevents the presence of large quantities of fluids and limits their mobility along joints and fracture systems. The igneous or metamorphic rocks may be considered as generally dry in comparison to permeable and porous sediments or heavily fractured rocks at shallow depth.

The problem how to tap this heat at depth was first attacked by physicists of the Los Alamos Scientific Laboratory (LASL) in New Mexico, USA in 1970. They proposed to drill boreholes to depth, to connect them by artificial fractures, and to circulate water between them thus cooling the hot rock in the vicinity of the fracture connection. The artificial fracture in the hot rock may be considered as a subsurface large-scale heat exchanger in the conventional sense. With respect to the assumption of the absence of natural fluids the idea was called the **Hot-Dry-Rock (HDR)** concept. The name remained until present although it was recognized that large quantities of fluids exist within the fracture network inspite of high pressure. Assuming that the technical problems to generate a sufficiently large fracture heat exchanger at depth is solved, the cooling of one cubic kilometre of rock by 100°C allows to operate a geothermal power plant of 30 MW (e) capacity for a period of 30 years. Speculating further, the concept may be developed into heat mining by cooling one cubic kilometre of rock after the other at one site. Calculations show that such systems can be operated economically and the environmental risks are estimated as minimal (Rummel and Kappelmeyer, 1993).

Hot-Dry-Rock research at the LASL has continued over almost 15 years with participation of Germany and Japan. In Europe, various complementary research was conducted in Germany, France, and England. In Japan, both the HDR projects at Hijiori and Ogachi started in the 1980s, Australia presently is preparing a new project. In Europe, the national research efforts were bundled in the European HDR research project at Soultz in the Upper Rhine Valley which will be developed into the first commercial industrial pilot project to produce electricity (see Klee, this volume). This pilot project may become a prototype for electricity production in small power plants of 10 to 20 MW (e) in the Upper Rhine Graben from Basel in the south, to Frankfurt in the north where temperatures of 200°C exist at 5 km depth, and due to its dense population and high industrialization consumers exist also for excess heat consumption. The technology developed within such pilot projects may be transferred for enhancement of existing geothermal systems with declining production (DOE 1999) and for large-scale heat production systems for adequate consumers. Presently the author prepares a geothermal project to supply the Ruhr University Bochum with heat from its own deep underground. The location is characterized with a rather normal geothermal gradient, i.e. 120°C at 3 to 4 km depth and fractured sandstones and shales of Upper Carbonian age. A heat production of 10 MW (th) from a borehole doublet is envisaged which could supply the base load heat demand throughout the year (Kattenstein et al., 2002).

9.4   GEOTHERMAL RESOURCES AND RESERVES

As shown in the previous sections the geothermal potential of the upper earth's crust is impressive and its present use in some few geothermal provinces with favourable geological boundary conditions is significant. However, it must be recognized that exploitation from developed geothermal systems is limited and only a small fraction of the totally stored geothermal energy can be technically exploited and economically used in the future even under the consideration of rapid development of new and advanced drilling and stimulation technologies.

Scenarios for the possible future use of crustal geothermal energy can be drawn from the McKelvy diagram (Fig. 9.3). In this diagram, the vertical axis is depth or the feasibility for technical/economic exploitation, the horizontal axis characterizes the status of resource exploration (identified, unknown, proven, probable existing, possible, in principal accessible, and economically feasible at some future time). Therefore, one may differentiate between the **accessible resource base** (ARB) which is the energy potential to the depth of present drilling capacity (7 km), and the **resource** which can presently be exploited and economically be used. The estimates on the resources is based on geological and geophysical data such as:

- dimension and depth of aquifers,
- genesis and properties of rock formations,
- salinity and chemistry of aquifer fluids,
- temperature and stress regime which controls fluid mobility within deep fracture networks.

Finally, the geothermal **reserves** are the resources which today can be used economically in competition on the energy market. Major factors for the estimation of reserves are exploration, drilling, and stimulation, and

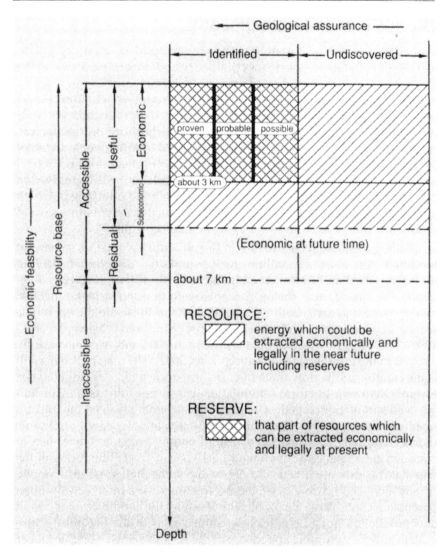

Fig. 9.3 McKelvey diagram defining geothermal resources and reserves as a function of economic feasibility

exploitation costs. The reserves are certain if sufficient reliable data from drillholes exist for the underground, they are probable if reservoir data are based on geological and geophysical explorations only, and are possible if geological imaging is the only data base (Muffler and Cataldi, 1978).

9.5 HOT-DRY-ROCK TECHNOLOGY

The HDR concept is based on the assumption that deep crystalline basement rock formations are practically dry and impermeable due to the high pressure of the overburden layers. Therefore, it was proposed to induce artificial fractures as heat exchangers through which fluid (water) is forced to flow from an injection borehole to the production borehole (Fig. 9.4). The injected water is heated from the hot rock on the fracture surfaces. In order to produce electricity of say 20 MW (e) over a period of 20 years, this requires fracture surfaces of several square kilometres, downhole rock temperatures of 200°C, and circulation with a production rate of more than 100 litres per second, and minimum pumping power to maintain circulation. In practice, the fluid production may require several production boreholes, and water losses should be negligible.

Drilling technology as existing in the oil industry must be somewhat modified for hard crystalline rock, similarly, directional drilling techniques exist and these can easily be used for drilling targets to 5 km depth. For casing, new casing concepts were developed using metallic casing packers to carry both the casing load and to isolate the open-hole section in which the artificial fractures must be induced (Hegemann et al., 1999). The concept was patented by MeSy in 2000 and was successfully applied in the European HDR project (Klee, 2002). The major problem still is the creation of the downhole fracture heat exchanger. The original idea to apply hydraulic fracturing stimulation was successfully used in oil- and gas-well stimulation, together with fracture propping to keep the fracture open for low hydraulic impedance fluid circulation was not applicable (Baria et al., 1999). Stimulation of a fluid path between boreholes had to consider the existence of the fracture network in crystalline rock and the fact that fractures are practically closed due to the high pressure at depth. In addition, only a few pre-existing fractures may intersect the open borehole section. Thus, the problem is two-fold: the boreholes for injection and production must be efficiently connected with the natural fracture system and the hydraulic impedance of the fractures must be reduced. For the first purpose, the author has proposed a hydraulic fracturing technology to zipper-like induce a fracture over most of the open-hole section of the borehole to improve the connectivity with the natural fractures. For the second purpose, it was recognized that purely tensile opening of favourably oriented fractures will not present permanent hydraulic paths if they are not kept open against the acting normal stress by proppings, and proppings may dissolve in the highly aggressive chemical environment. On the other hand, if fractures inclined with respect to tectonic stresses are opened they might shear and prevent closure after the stimulation operation. Shear on fractures is associated

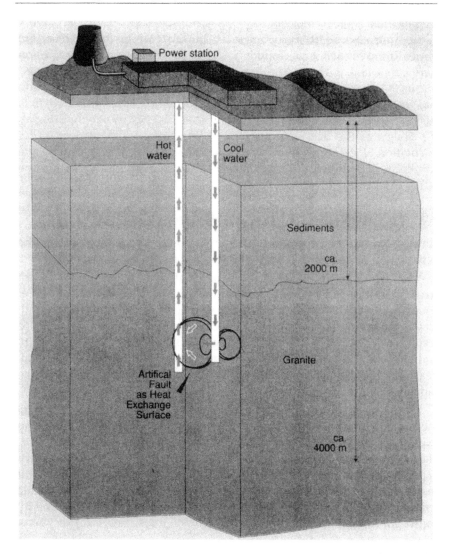

Fig. 9.4 Hot-Dry-Rock (HDR) concept for geothermal energy extraction from deep hot crystalline rocks

with induced seismicity which allows to locate and follow the effect of stimulation in space and time. Thus, rock mechanics has become an important tool in HDR reservoir engineering by determining the magnitude and orientation of the in-situ tectonic stresses at depth, to predict shear displacement on fractures, and to understand the hydraulic behaviour of fractures in terms of hydraulic valves.

One final aspect is the concern on major fluid losses during circulation. In the European HDR project in the fractured granitic basement of the Upper Rhine Graben it was found that the deep drillholes after stimulation were connected with a huge fluid reservoir existing in the fluid stored within the natural fracture network. On the other hand, the production rate was equal to the rate at which the produced fluid was reinjected, thus the loss rate was practically zero. This was clearly the result of optimizing the design of the drillhole network configuration with respect to the orientation of the tectonic stress regime: the drillholes are along a profile parallel to the direction of maximum horizontal stress and within the rather planar seismic cloud of induced seismicity.

9.6 ENVIRONMENT AND GEOTHERMAL ENERGY

Similar to the consumption of conventional energy resources the exploitation of geothermal heat is an interference in the environment.

The use of geothermal energy is associated with the production of large quantities of hot water/steam from porous aquifers or from HDR circulation systems. Most of this water will be reinjected into the ground. Nevertheless such operation may effect the underground water balance resulting in pore pressure changes, regional subsidence, microseismicity and emission of gasses, and production of solved minerals. In addition, heat may be released into the atmosphere and surface installation like pipelines will be a nuisance to nature.

Until present environmental risk analysis in Europe are conducted on the basis of the European Community standard No. 85/337/EEC. The exploitation of geothermal resources is subjected to this standard. However, neither authorities nor industrial users have the necessary experiences to apply this standard to geothermics with full responsibilities. Therefore, present enterprises require responsible considerations to develop geothermal energy towards a future energy source for mankind.

REFERENCES

Baria, R., Baumgärtner, J., Rummel, F., Pine, R.J., Sato, Y. 1999. HDR-HWR reservoirs: concepts, understanding and creation. Geothermics, 28, 4/5, 533-552, Pergamon.

DOE. Feb. 1999. Strategic roadmap for the enhanced geothermal systems research and development program. US Dept. of Energy, Office Geothermal Development.

Freeston, D.H. 1995. Direct uses of geothermal energy 1995 - Preliminary review. Proc. World Geotherm. Congr. Vol. 1, 15-25.

Hegemann, P., Klee, G., Rummel, F. 1999. Metal sealing packer technology in deep boreholes. Bull. d'Hydrogeologie, 17, 159-163, Neuchatel.

Huttrer, G.W. 1995. The status of world geothermal power production 1990-1994. Proc World Geotherm. Congr. Vol. 1, 3-14.

Klee, G. 2002. The European Hot-Dry-Rock project in the tectonic regime of the Upper Rhine graben. This volume.

Muffler, L.J.P., Cataldi, R. 1978. Methods for regional assessment of geothermal resources. Geothermics, 7, 53-90.

Rummel, F. 1987. Fracture mechanics approach to hydraulic fracturing stress measurements. In : Fracture Mechanics of Rocks, (ed. B. Atkinson), Academic Press, 217-239.

Rummel, F. 2002. Crustal stress derived from fluid injection tests in boreholes. In: In-Situ Characterization of Rocks (eds. V.M. Sharma and K.R. Saxena), 6, 205-244, Balkema.

Rummel, F., Kappelmeyer, O. 1993. Geothermal energy - Future energy source? Publ. CF Mueller Karlsruhe, p98.

Sekioka, M. and Toya, S. 1995. Country update report of geothermal direct uses in Japan. Proc. World Geoth. Congr. Vol. 1, 217-222.

Yamaguchi, F., Ohishi, K., and Esaki, Y. 1995. Current status of development of geothermal power generation in Japan. Proc. World Geoth. Congr. Vol. 1, 209-215.

10

The European Hot-Dry-Rock Project in the Tectonic Regime of the Upper Rhine Graben

Gerd Klee

ABSTRACT

Deep hot crustal rocks offer an almost unexhaustible energy resource. In 1970, physicists of the US Los Alamos Scientific Laboratory (LASL) proposed a concept, the so-called Hot-Dry-Rock (HDR) concept, to tap this geothermal energy source, by circulating water through an artificially created heat exchanger between adjacent deep boreholes. This concept was studied by various geoscience groups in Europe at different locations during the 80s. In 1987 the European HDR research efforts merged into the Scientific European HDR project at the location Soultz-sous-Foret in the Upper Rhine Graben near the boundary between France and Germany. It was possible to create a large heat exchanger between two 3.5 km deep boreholes approximate 500 m apart from each other by high pressure water injection (hydraulic fracturing). In 1997, a four-month circulation test demonstrated a geothermal energy capacity of approximately 10 MW at a downhole rock temperature of 160°C. The test, in particular, showed that the tectonic graben-type stress field is the controlling factor for water circulation on induced and stimulated fractures within the granitic basement rock. Since 2000, the system was deepened to a depth of 5 km with a rock temperature of 200°C. This system consisting of two production and one re-injection wells, operated by a European Industrial Consortium (EIEG), will be developed into a first HDR pilot plant to produce 5 to 10 MW electricity in 2006.

10.1 INTRODUCTION

The use of geothermal energy today is limited to hot water or steam deposits located in areas with specific geological and tectonic conditions which generally exist along plate boundaries (e.g. California, Japan, and in Mediterranean countries). By far the largest heat resource, however, exists in the hot basement of the continental crust, which could be tapped almost everywhere and anytime. This was first recognized by scientists from the Los Alamos Scientific Laboratory, New Mexico, USA, and led to the formulation of the Hot-Dry-Rock concept in 1970. Other countries like Germany, France, England, Sweden, and Japan contributed to the LASL project, but also developed their own national HDR research within their countries (e.g. Rummel and Kappelmeyer, 1993).

In 1986-87, scientists from Germany, France, and England combined their research efforts, agreed on a common test site in the Upper Rhine valley, and started the European HDR-project. The selected research site is situated at Soultz-sous-Forêts, about 50 km north of Strasbourg (Alsace, France) near the western margin of the Upper Rhine Graben, the most prominent present-day tectonic unit in Europe, north of the Alps (Fig. 10.1). The area is characterized by moderate seismicity, a thin continental crust (less than 25 km) and a well-known geothermal anomaly, with high heat flow of about 150 mW/m^2 and geothermal gradients of up to 100°C/km in the uppermost sediments to about 1.4 km depth and about 30°C/km in the underlying granitic basement. The anomaly is a result of deep water convection through a dense fracture network in the sediments and was identified from numerous measurements in the former local Pechelbronn oil-field.

The process of the Upper Rhine valley development was controlled by the opening of a pre-existing fault zone under compression parallel to the NNE-SSW orientation of the graben (parallel to the paleo-stress direction), crustal rifting due to the development of a mantle diapir in the southern graben and the evolution of the present-day stress regime. The earthquake focal mechanism data for the Upper Rhine valley indicate mainly both, strike-slip and normal faulting stress conditions with a consistent maximum horizontal stress direction of NW-SE, which is in agreement with borehole breakout and hydrofrac stress data from boreholes in the central part of the Rhine valley and in the Black Forest (Müller et al., 1992).

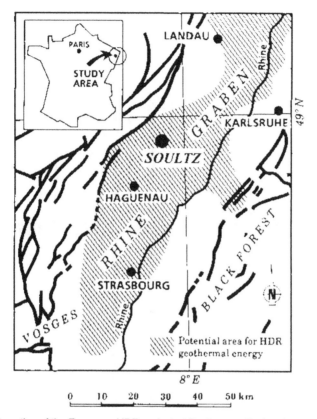

Fig. 10.1 Location of the European HDR project at Soultz-sous-Forêts, France

10.2 PROJECT DEVELOPMENT DURING 1987-2000

Due to regulations of the funding agencies (European Commission, France, Germany) the project was developed in steps of 2-3 years duration since 1987. During this time the project has been continuously reinforced in terms of infrastructure at the site, a borehole network, and massive stimulation operations.

During the initial project phase between 1987 and 1988, borehole GPK-1 (Fig. 10.2) was drilled to 2,000 m depth with a bottom hole temperature of 140°C. A number of small-scale hydraulic tests and geophysical measurements were conducted to investigate the rock mass conditions to 2 km depth (Bresee, 1992).

During 1989-1991, the first seismic network was installed in three recovered old oil wells (Fig. 10.2, borehole nos. 4550, 4601, and 4616). Borehole GPK-1 was stimulated with 7 and 15 l/s to connect the bottom section of the borehole with a permeable fault system. These tests already

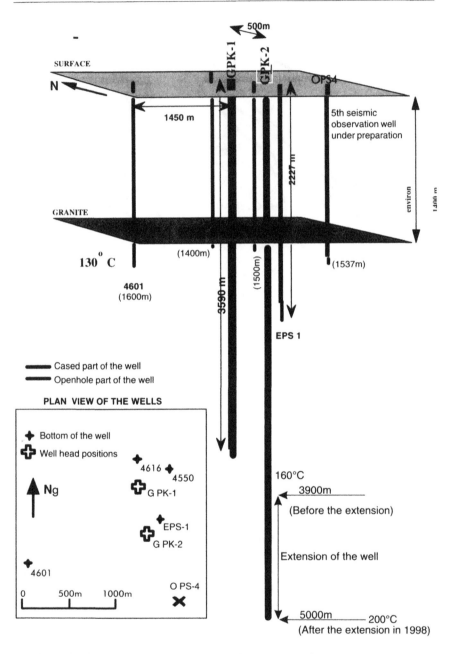

Fig. 10.2 Borehole network at the European HDR project site Soultz-sous-Forêt

showed a proportionality between stimulation flow rate and production rate, one of the major findings of the project which contributed to the planning of subsequent stimulation operations. An attempt was also made to drill the continuous cored borehole EPS-1 to 3,500 m depth. However, the well was terminated at 2,227 m due to drilling problems. Nevertheless, extensive investigations on core samples and cuttings of the borehole yield important data on the mechanics, joint network, and mineralogy, and provided the basis for subsequent interpretations of geophysical logs. Today, borehole EPS-1 is also used for micro-seismic monitoring (Fig. 10.2).

During 1992-1993, the existing borehole GPK-1 was deepened from 2,000 m to 3,590 m depth with a bottom hole temperature of 159°C. Various geophysical measurements were conducted during and after drilling. Following the drilling of the well, two massive stimulation tests with a maximum flow rate of 50 l/s were carried out in the open-hole section of borehole GPK-1 below 2,850 m depth. During these tests, a 2 km^2 large fracture system was stimulated which was well connected with a highly permeable fault system (Baria et al., 1995). Subsequently, the fracture system was extensively investigated with the highlight of the first production test in June 1994 (Jung et al., 1995).

The large body of existing information on in-situ stress, temperature, joint network, and seismicity was than used to target the drilling of the second deep borehole GPK-2 to 3,890 m depth, located approximately 450 m south of borehole GPK-1 (Fig. 10.2). After the well was completed in early 1995 with a bottom hole temperature of 175°C, a massive stimulation test was conducted in borehole GPK-2 below 3,200 m depth, where a fracture system of approximately 1 km^2 was created and connected to the stimulated reservoir of borehole GPK-1. At the end of this phase, a 10-days circulation test between the two deep boreholes was carried out. Since several hydraulic tests showed that the fracture system at Soultz is hydraulically open at its periphery, it was concluded that the circulation could not be operated at a fluid pressure above hydrostatic as this would cause high fluid losses. Therefore, a downhole pump was used with great success in the production borehole GPK-1. The test showed a stable equilibrated flow of 20 l/s at a temperature of 135°C and a thermal power of 8 MW (th). Since the produced fluid was re-injected, no fluid losses occurred (Baumgärtner et al., 1996).

In order to demonstrate that such a circulation can be maintained over a longer period, a circulation test of 4 months duration was conducted during summer and autumn 1997. Prior to this test, borehole GPK-2 was re-stimulated with flow rates up to 78 l/s to reduce the flow resistance at the fracture inlet which increased during the first circulation test (Gérard et al., 1997). The set-up of the circulation experiment is shown in Fig. 10.3, where the produced brine is kept within a fully closed loop on surface to prevent scaling and corrosion. A submersible pump was installed in the

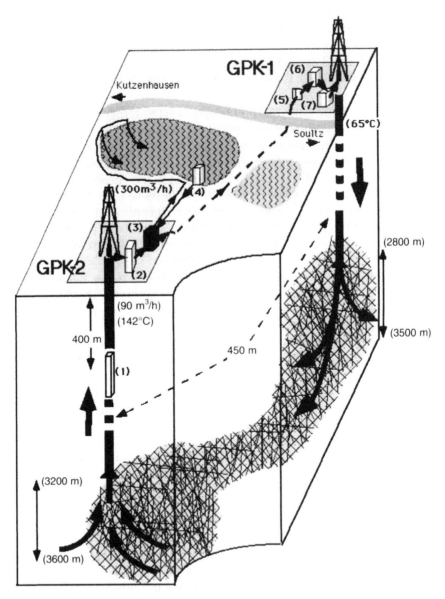

Fig. 10.3 Circulation tests between borehole GPK-1 and GPK-2 in 1997 (Gérard et al., 1999). (1) Submersible pump, (2) Pre-filter, (3) Heat-exchanger, (4) Pumps for cooling loop, (5) Corrosion test chambers, (6) Filter battery, (7) Re-injection pump.

production borehole GPK-2 at about 400 m depth. The produced brine was pre-filtered before entering the cooling loop consisting of a heat-exchanger to extract a large portion of the recovered heat and an artificial lagoon. The cooled water was then transmitted via a composite line to the platform of borehole GPK-1. Before re-injection, the brine was again filtered. The circulation-system infrastructure installed was automated and fully instrumented with online data monitoring.

Neglecting some initial problems due to unexpected power failures and short interruptions for equipment maintenance and cleaning, the experiment was continued for four months without major technical problems and noticeable environmental impact (Fig. 10.4). The circulation was maintained without adding any additional fluid, production and re-injection were fully balanced. The produced flow increased stepwise from 21 l/s to 25 l/s (90 tons/hour). By the end of the test, 244,000 tons of brine were recovered with a temperature of 142°C. The re-injection pressure in borehole GPK-1 decreased from about 4.5 MPa to 2 MPa after the injection of an anti-scaling agent was cancelled (the anti-scaling agent may have caused some skin effect in the near well-bore area). The thermal energy available at the heat exchanger was in the order of 10-11 MW (th) (based on a re-injection temperature of 40°C for space heating installations). On the other hand, the electric energy necessary to maintain the circulation was only about 250 kW (Baumgärtner et al., 1998).

In the later stage of the project during 1998-2000, the accuracy of the micro-seismic monitoring system has been improved by drilling the additional observation borehole OPS-4 located in the south of the network (Fig. 10.2). In order to investigate the conditions in the temperature region of 200°C (sufficient for electricity generation), borehole GPK-2 was deepened to 5,084 m depth. The operation was accompanied by the development of new casing packer elements based on inflatable metal shells which were successfully integrated into the completion of the well (Hegemann et al., 1999). Finally, the new reservoir below the casing shoe at 4,430 m depth was stimulated with flow rates up to 50 l/s. Presently, two further deep boreholes to 5,000 m depth were completed to develop a first HDR pilot plant for electrical power generation of approximately 5 MW_{el}.

10.3 IN-SITU HYDROFRAC TESTS AND STRESS REGIME

The efficiency of the heat exchange by fluid circulation between deep boreholes strongly depends upon the hydraulic impedance of the fluid flow path. Besides permeability, temperature, and chemical reactions between the circulating fluid and the crystalline rock-mass, the hydraulic properties at depth are mainly controlled by the in-situ stress regime.

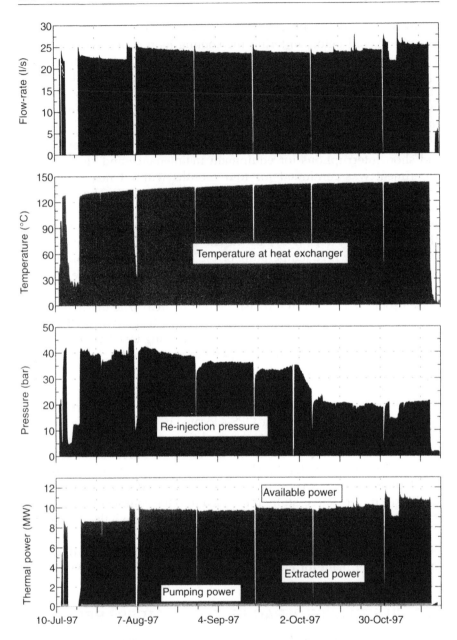

Fig. 10.4   Major parameters of the 1997 circulation experiment (Baumgärtner et al., 1998)

Therefore, since the early beginning of the Soultz HDR project, the tectonic situation in the Upper Rhine valley has been extensively investigated by various hydraulic fracturing stress measurements down to 3.5 km depth (Rummel and Baumgärtner, 1991; Klee and Rummel, 1993).

- During the first project phase in late 1988, a total of eight hydrofrac/hydraulic injection tests were carried out in borehole GPK-1 between 1,458 m and 2,000 m depth in conjunction with the small-scale hydraulic test program.
- Two hydrofrac tests were conducted in borehole EPS-1 at about 2,200 m depth in late 1991.
- After deepening of borehole GPK-1 to almost 3.6 km depth, two further tests were carried out at 3,315 m and 3,506 m depth in 1992.

Although successfully completed, the first test series in borehole GPK-1 was characterized by several technical problems caused by using conventional (rubber-based) packer technology in the hostile downhole environment (temperatures up to 140°C and high gas and salt content of the borehole fluid). Therefore, the later tests in borehole EPS-1 and GPK-1 were conducted using ductile metallic packers as part of a wireline hydrofrac system (Klee and Rummel, 1993).

10.3.1 Aluminium Straddle Packer Tool

For the first approach, aluminium was selected as packer material on account of its high ductility, good machining properties, the low cost. After several tests with laboratory models, aluminium straddle packer tools for borehole diameter of 4, 5-7/8, 6-1/4, and 8-1/2 inch were designed. A schematic view showing the essential details of the system is given in Fig. 10.5. The major characteristics are as follows.

The packer elements consist of pure (soft) aluminium (Al 99.5 %) which allows a maximum deformation of 25 % at room temperature. However, the outer diameters and the wall thickness are designed in such a way to accommodate the borehole diameter by approximately 15 % of lateral deformation and differential pressures of about 25 MPa. To guarantee sealing, the outer surface of the aluminium packers is furnished with high temperature O-ring seals. The two soft aluminium packer elements are connected with threads to the injection interval part and the end pieces, both made from high strength aluminium alloy (ERGAL 55).

For separate pressurization of both, the packer elements and the injection interval, an inner stainless steel mandrel is used. This mandrel contains high temperature O-rings as seals against the aluminium shells and deep borings for hydraulic connection to the packer inflation sections and the injection interval. The aluminium parts of the arrangement are

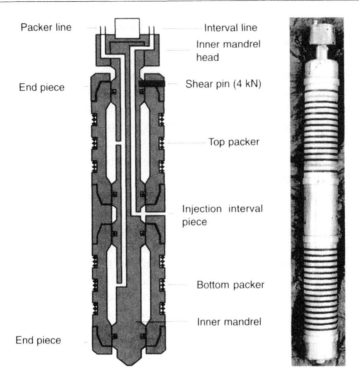

Packer line — Interval line
Inner mandrel head
End piece — Shear pin (4 kN)
Top packer
Injection interval piece
Bottom packer
Inner mandrel
End piece —

Fig. 10.5 Schematic diagram and photograph of the aluminium straddle packer tool

fixed on the inner mandrel with a 4 kN shear pin. This design enables all steel parts of the tool to be recovered after completion of the test, while the aluminium parts remain in the borehole. The aluminium can be drilled out by adequate drilling procedures.

Besides the application of the metal packer technology for hydro-fracturing at great depth and high temperature, the technology has been investigated with promising results for permanent borehole sealing in nuclear waste storage projects and for the reliable casing cementation/ anchoring at severe downhole conditions (Hegemann et al., 1999).

10.3.2   In-Situ Stress Data

The results of the stress measurements yield:

- An orientation of the acting major horizontal stress S_H of N-S to NNW-SSE which is in accordance with existing stress data for Central Europe (Fig. 10.6). The stress direction was verified by the spatial distribution of thousands of induced seismic events during the massive stimulation tests in borehole GPK-1 and GPK-2.

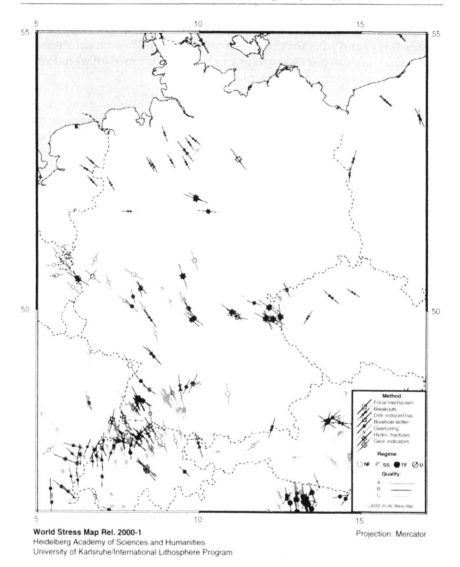

World Stress Map Rel. 2000-1
Heidelberg Academy of Sciences and Humanities
University of Karlsruhe/International Lithosphere Program

Projection: Mercator

Fig. 10.6 Orientation of the maximum horizontal stress S_H in Central Europe (Müller et al., 2000)

- A stress regime with notably low horizontal stresses, typical for the tectonic situation in the Upper Rhine graben (e.g. Klee and Rummel, 1999):

$$S_h = 15.7 + 0.0149 \times (z - 1458)$$
$$S_H = 23.5 + 0.0337 \times (z - 1458)$$

where S_h and S_H are the minimum and maximum horizontal principal stresses (MPa) and z is the depth below surface (m). The vertical stress S_v can be calculated from the weight of the overburden with given rock density ($\rho = 2.66 \text{ g/cm}^3$ in the granite):

$$S_v = 33.1 + 0.0261 \times (z - 1377)$$

The stress profiles are shown in Fig. 10.7. As demonstrated by the stimulation tests in borehole GPK-2 (95JUN16, 96SEP18), the pressure for massive fluid injection into favourable oriented joints is controlled mainly by the minimum horizontal stress component S_h. Therefore, reliable stress data can be used for the technical planning of injection tests at great depth.

A stability analysis on the basis of a simple friction law $|\tau_c| = \mu \cdot \bar{\sigma} = \mu \cdot (\sigma - k \cdot P_o)$ where τ_c is the critical shear stress, σ the effective normal stress, P_o the local hydrostatic pressure, and μ the friction coefficient and the stress-profile equations leads to an estimation of the critical differential stress at which sliding on favourable faults or joints

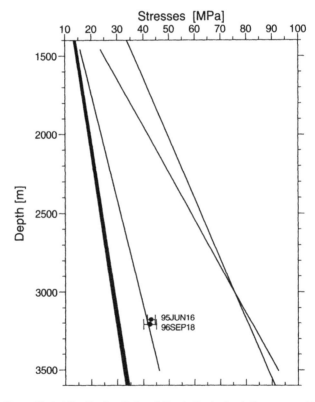

Fig. 10.7 Stress-filed at the Soultz site in relation to the hydrostatic pressure P_{hyd}. 95JUN16 and 96SEP18 marks the injection pressure for massive fluid injection in borehole GPK-2.

might occur. k is the pore-pressure ratio with respect to hydrostatic conditions (k = 0: no pore pressure, k > 1: over-hydrostatic conditions). In Fig. 10.8, a comparison of calculated critical differential stresses with the experimental determined and linearly extrapolated results to 5 km depth is shown. The calculations were carried out by using a friction coefficient of μ = 0.85 for a normal faulting stress regime (similar results were obtained for a strike-slip faulting stress regime below 3 km depth). The analysis demonstrates that minor reservoir fluid pressure variations (pore pressure values slightly higher than the hydrostatic pressure) already will induce micro-seismicity and will release the stored elastic energy within the reservoir.

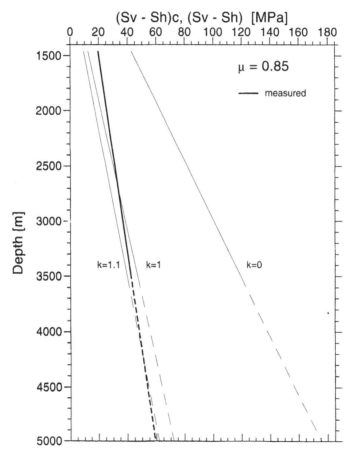

Fig. 10.8 Comparison between critical differential stresses calculated for different pore pressure ratios, a friction coefficient of 0.85 and a normal faulting stress regime with the in-situ measured and extrapolated data (dashed lines)

10.4 CONCLUSIONS

During 15 years of research within the European HDR project at Soultz-sous-Forêts, the fracture network in the Upper Rhine graben has been explored down to 5 km depth with a temperature of more than 200°C. Besides the large geothermal anomaly, one of the secrets for the successful project development was the detailed consideration of the tectonic stress situation at depth, measured by various hydraulic fracturing tests to a depth of 3.5 km using metallic packers suitable for the hot and chemically aggressive downhole environment in the Soultz boreholes. The results yield an in-situ stress direction of NWW-SEE in accordance with existing stress data for Central Europe, and an in-situ stress regime with notably low horizontal stresses, typical for the normal faulting graben tectonics which offers favourable conditions for massive fluid circulation tests.

Later, during various stimulation and hydraulic tests including a circulation experiment of four months duration, the favourable conditions in a graben structure were verified. The four-months circulation test demonstrated that a circulation loop can be maintained with flow rates up to 90 tons per hour and a fluid temperature of more than 140°C, between two boreholes of 450 m distance without any water losses and requiring only 250 kW_{el} pumping power compared to a thermal output of about 11 MW_{th}.

Future work will include the design and construction of a pre-industrial prototype enabling forced fluid circulation between a central injection borehole and two deviated productions wells at depth with temperatures of 200°C. Such a scientific pilot plant could produce approximately 50 MW_{th} over a period of several years and enables electric power generation of approximately 5 MW_{el}.

ACKNOWLEDGEMENT

MeSy's active participation in the Soultz project since 1986, in particular the conduction of the in-situ stress measurements as well as the development of the metal packer technology, were funded by the German Ministry of Research within several research contracts. The successful cooperation with the project coordinator Socomine and all other project partners is acknowledged.

REFERENCES

Baria, R., Garnish, J. Baumgärtner, J. Gérard, A. and Jung, R. 1995. Recent Developments in the European HDR Research Programme at Soultz-sous-Forêts (France). Proceedings of the World Geothermal Congress, Florence, 4:2631-2637.

Baumgärtner, J., Jung, R., Gérard, A., Baria, R., and Garnish, J. 1996. The European HDR project at Soultz-sous-Forêts: stimulation of the second deep well and first circulation experiments. Proceedings of the 21st Workshop on Geothermal Reservoir Engineering, Stanford: 267-274.

Baumgärtner, J., Gérard, A., Baria, R., Jung, R., Tran-Viet, T., Gandy, T., Aquilina L., and Garnish, J. 1998. Circulating the HDR reservoir at Soultz: maintaining production and injection flow in complete balance. Proceedings of the 23rd Workshop on Geothermal Reservoir Engineering, Stanford.

Bresee, J.C. 1992 (edition). Geothermal energy in Europe. The Soultz Hot Dry Rock project. Gordon and Breach Science Publication.

Gérard, A., Baumgärtner, J., and Baria, R. 1997. An attempt towards a conceptual model derived from 1993-1996 hydraulic operations at Soultz. Proceedings of NEDO International Geothermal Symposium, Sendai, 2:329-341.

Hegemann, P., Klee G., and Rummel, F. 1999. Metal sealing packer technology in deep boreholes. Bull. d'Hydrogéologie, Université de Neuchâtel 17:159-163.

Jung, R., Willis-Richard, J. Nicholls, J. Bertozzi A., and Heinemann, B. 1995. Evaluation of hydraulic tests at Soultz-sous-Forêts, European HDR-site. Proceedings of the World Geothermal Congress, Florence 4:2671-2676.

Klee, G. and Rummel, F. 1993. Hydrofrac stress data for the European HDR research project test site Soultz-sous-Forêts. Int. J. Rock Mech. Min. Sci. & Geomech. Abstr. 30, 7:973-976.

Klee, G. and Rummel, F. 1999. Stress regime in the Rhinegraben basement and in the surrounding tectonic units. Bull. d'Hydrogéologie, Université de Neuchâtel 17:135-142.

Müller, B., Zoback, M.L., Fuchs, K., Mastin, L., Gregersen, S., Pavoni, N., Stephansson O., and Ljunggren, C. 1992. Regional pattern of tectonic stress in Europe. J. Geophys. Res. 97 (B8):11783-11803.

Müller, B., Reinecker J., Heidbach O., and Fuchs. K. 2000. The 2000 release of the World Stress Map (available online at www.world-stress-map.org).

Rummel, F. and Baumgärtner, J. 1991. Hydraulic fracturing stress measurements in the GPK-1 borehole, Soultz-sous-Forêts. Geotherm. Sci. & Tech., 3 (1-4):119-148.

Rummel, F. and Kappelmeyer, O. 1993 (eds). Geothermal energy - future energy source. Verlag C.F. Müller.

11

Underground Storage of Natural Gas

Georg Boor and **Michael Krieter**

ABSTRACT

In order to balance alternating annual gas supply and demand or to create strategic gas reserves, different types of underground gas storage (UGS) facilities, such as gas caverns, depleted gas/oil fields and aquifers, etc., are used by national and international gas suppliers. The following chapter describes all types of UGS facilities more or less in detail, depending on their significance. At present, about 600 UGS facilities exist worldwide, 90% being porous rock storage facilities and the other 10% involving cavern storage facilities. Worldwide storage capacity and distribution is presented and tabulated in the last part of this chapter.

11.1 INTRODUCTION

The gas fields from which natural gas is brought to centres of consumption are usually situated in remote locations. To take this gas to the market, an extensive system of pipelines has been built around the world. Europe, for example, has its main sources of natural gas in Russia, the North Sea, the Netherland, and in Algeria. Because of the long distances involved, the pipeline systems have to be sized to carry enormous volumes.

Building gas transmission systems requires substantial investments. These investments and operating costs have to be recouped through the gas transmission business. For this reason, every pipeline operator will first of all try to achieve full capacity utilization on his system before building any additional capacity which is only needed for transmission peaks. Optimizing transmission system operation is therefore basically about trying to operate the system as continuously and as close to the maximum system capacity as possible.

11.2 BALANCING SUPPLY AND DEMAND

Natural gas supply and demand structures are not the same, while producers and pipeline operators try to run their facilities as constantly as possible during the course of a day/year, gas consumption varies depending on the time of the day and the season (summer/winter).

The changing daily or monthly gas flows show a distinct demand curve over the course of a year, which is usually shown on a daily or monthly basis. Fig. 11.1 is an example depicting typical annual demand curve of a country.

During the heating season from January to March and November to December, consumption is between twice and three times as high as in summer. This ratio is mainly dependent on the consumption structure. While power stations and industry require a more or less constant flow of gas throughout the year, consumption by small commercial and residential users is closely linked to the temperatures. So the ratio between maximum winter demand and minimum consumption in summer depends on the share of each individual consumer sector in overall gas consumption. Normally, this ratio is around 2 to 4, but in extreme circumstances it can be higher if producers reduce their supplies within the agreed limits during the winter and consumers are more or less fully supplied from underground storage facilities. Fig. 11.2 show the determination of a annual storage demand.

The graph shows the quantities under the supply agreement including the agreed flexibility and the maximum daily quantity needed (green line). The daily quantities planned are represented by the red line. The curves show that the contract quantities cover gas demand on about 265 days. For the remaining 100 days of the year the demand exceeds the supply volume. For this period, additional gas will have to be withdrawn from storage.

The required storage injection capacity is determined in more or less the same way, except that for this purpose, the minimum daily quantities under the supply agreement are plotted together with the injection capacity needed for refilling the storage facility in about 200 days. The

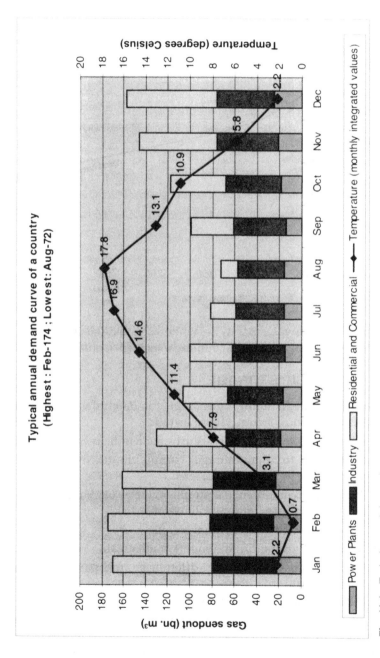

Fig. 11.1 Typical annual demand curve

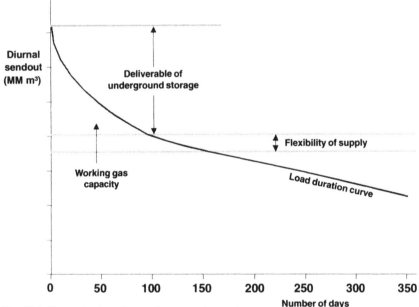

Fig. 11.2 Determination of annual storage demand

maximum withdrawal capacity in winter and the injection capacity in summer define the working gas volume to be kept available.

11.3 STRATEGIC STORAGE

The need for strategic natural gas storage is determined by a range of economic and political factors. These include:

- The need to ensure supply security even under extreme conditions.
- Being able to maintain gas supply in the event of supply disruptions for technical reasons.
- Being able to maintain gas supply in the event of reduced supplies or supply failures for political reasons.
- Gas production system optimisation.
- Third-party storage in the country and abroad.
- Use of short-term price differences for gas (spot market, speculation).
- Administrative requirements.

These factors can actually lead to strategic storage of gas vary quite significantly from case to case. In most cases, subjective reasons and experiences also play an major role. However, the most important factor for the economic viability of gas storage is the ratio between the investments and the operating costs on the one hand and the frequency of strategic capacity use on the other.

Strategic storage facilities are built along or close to the main gas transmission routes. Preventially porous rock storage facilities are used for strategic storage because the specific costs for large-volume cavities are well above those of porous rock facilities. Porous rock facilities which, when used for strategic purposes have working gas volumes of 2 billion m³ and more, are usually built in depleted or partly depleted natural gas reservoirs. These facilities are then operated from the pipeline grid control centre.

Examples of underground storage facilities in Europe only serving strategic purposes include the Bilce Volisk facility in the Ukraine (back-up for Russian gas supplies to western and southern Europe), Etzel in Germany (back-up for Norwegian gas supplies to Ruhrgas AG) and Norg as well as Grijpsperk in the Netherlands (used for storing large capacities to improve market entry opportunities arising from competition and deregulation in Europe). In France, the level of strategic reserves is prescribed by law. It requires average gas consumption to be secured by additional capacities for a period of 110 days.

11.4 DIFFERENT TYPES OF UNDERGROUND STORAGE

Underground natural gas storage in porous reservoirs is the most convenient means of storing gas to meet the needs of the market during the peak winter demand period. The most effective means of storage is naturally in depleted reservoirs. In the absence of such structures, however, natural gas can be stored in aquifers and in salt cavities. Examples are also available of natural gas storage in disused mines. The different storage methods used from one country to another depend largely on the geological formations available. They are presented in Fig. 11.3 and Fig. 11.4.

In a storage facility, only part of the total volume of the gas stored is available (working gas), while the remainder is the cushion gas (or base gas). These two quantities are approximately equal. The ratio between these two figures depends on the specific conditions of the reservoir. A storage facility is characterized by:

- the working gas volume,
- the maximum send-out capacity

The ratio of these two figures serves to determine the maximum daily withdrawal time. In practice, the maximum daily production capacity is limited by the number of wells and the capacity of the surface equipment. The maximum withdrawal capacity in winter (or adjustment) is limited by

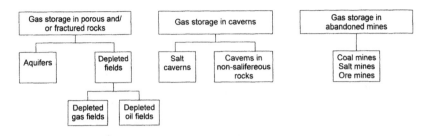

Fig. 11.3   Different types of underground storage

the volume of gas stored in the beginning of winter and consequently depends on the reservoir characteristics.

Based on these criteria, two operating types of storage facility can be distinguished.

- With a working capacity of 10 to 20 days of peak daily withdrawal (peak storage). The installations built in salt or disused mine cavities correspond to this type of facility, and have to satisfy peak demand. They offer the advantage of productivity per well that is two to four times higher than that of porous storage facilities.

- With a working capacity of 50 to 100 days of peak daily withdrawal (seasonal storage). Storage facilities in aquifers and depleted reservoirs belong to this class. They adequately meet the need for seasonal adjustment, because the working volume that can be stored is larger than in mined cavities (ratio of 1 to 10).

Underground storage sites represent the most effective means of balancing gas supply and demand, naturally combined with the other means available to the gas industry (interruptible contracts, supply flexibility). However, the gas industry has developed other methods, particularly peak-shaving units, which can supply gas at high and rapid rates in the cold season and over a short interval. LNG storage is also integrated with the LNG import terminals.

11.4.1   Gas Storage in Porous And/Or Fractured Rocks

Gas storage in porous and/or fractured rocks was introduced in the early 20th century and has got a worldwide use now. The gas is stored in the tiny pores of a connected pore space of the rocks and/or in the fracture systems of the rocks.

There are two main types of gas storage in porous rocks:

- depleted reservoirs and
- aquifers.

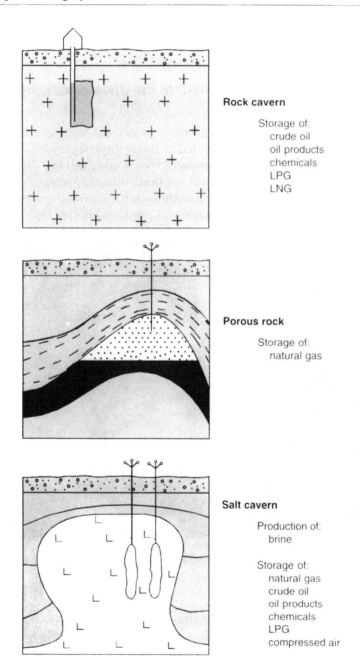

Rock cavern

Storage of:
crude oil
oil products
chemicals
LPG
LNG

Porous rock

Storage of:
natural gas

Salt cavern

Production of:
brine

Storage of:
natural gas
crude oil
oil products
chemicals
LPG
compressed air

Fig. 11.4 Schematic presentation of main types of underground storage

Besides the economical and geographical conditions some geological requirements are important, both for depleted reservoirs and for aquifers, when used as an underground gas storage.

11.4.2 Subsurface Rock Suited for the Underground Storage of Gas

The presumed storage horizon must have a sufficient thickness and good reservoir properties (e.g. porosity and permeability), layered in a suited depth (common depths are between 500 and 2,000 m). Generally sandstones or unconsolidated sands are best suited as storage horizons, but also fractured limestones or dolomites are common. A section of a microscope photo of a porous sandstone is presented in Fig. 11.5.

Fig.11.5 Microscope photo of a porous sandstone

11.4.3 Sealing Horizon

To prevent any gas migration to shallower horizons or even to the surface the presumed storage horizon must be covered by a sealing horizon of sufficient thickness and gas-tight properties even at higher pressures of future storage operations. The main types of sealing caprocks are generally clay, marl, or rock salt. Less suited are limestones, dolomites, anhydrites, or tight quarzitic sandstones because of possible fractures and joints.

11.4.4 Subsurface Trap of Sufficient Size and Geological Integrity to its Edges (Closure)

There are structural, facial, and stratigraphic traps, very often combined with each other. Structural traps, subdivided into monoclines and anticlines, are the most important type of traps and generally easier to define and to prove by seismic investigations. Anticlines are characterized by an all around culmination of the subsurface strata (4-dip closure), whereas monoclines are not updomed to all directions, but limited by tectonic faults. These faults must have a tight fault plane, preventing any communication between the separated blocks, even of juxtapositions of porous rocks.

Facial traps are characterized by a lateral change of the porous and permeable reservoir rocks to tight horizons, generally clay or marl (shale-out of the reservoir rock). Also old reefs, surrounded by impermeable rocks, can be suited for underground storages. Generally however, facial traps are difficult to prove by standard exploration methods.

Stratigraphic traps are characterized by unconformities, truncating the reservoir rocks. The onlap horizon must be a tight clay or marl. The closure area and closure height (altitude difference between the top of the trap and the spill-point) must be suited for the gas storage and must enclose an area of sufficient size. The different types of traps are presented in Fig. 11.6.

All these requirements have to be explored very carefully before using aquifer structures for underground storage purposes by seismic survey, drilling works, laboratory tests on drilling cores, by hydraulic tests (e.g. interference tests between individual wells or even cross-over tests between individual wells and different horizons), and by model simulations.

When converting a depleted reservoir, however, most of these data and requirements are already well known by the exploration and exploitation phase of the reservoir, so only additional investigations of minor volumes are to be necessary (e.g. the threshold pressure tests of the caprock, limiting the future maximum injection pressure and model simulations).

11.4.5 Depleted Reservoirs

Gas storage in depleted reservoirs is the most widespread method in the world and the most economical. Most of these are depleted gas reservoirs, and a few depleted oil reservoirs are also employed for the purpose.

The principle of this type of storage facility is simple, because the field formerly contained gas or oil. It therefore also fulfills, in most cases, the permeability and porosity conditions required for storage. However, before the facility is developed, the depleted field must be studied to be

Anticlinal structure

Fault

Facial trap

Stratigraphic trap

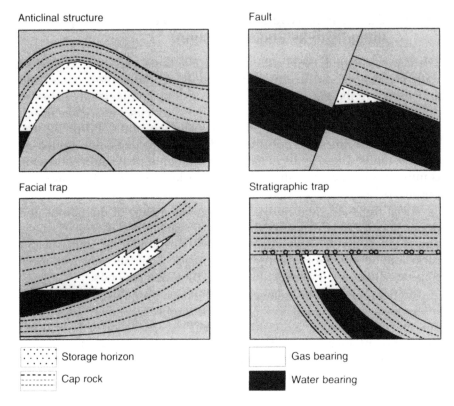

```
:::::::::  Storage horizon          Gas bearing

---------  Cap rock                 Water bearing
---------
```

Fig. 11.6 Different types of traps.

sure that it can meet the production targets (high withdrawal rates over short periods) and that the impermeable formation above the storage zone, or cap rock, is sufficiently gas tight. Only minimum risk is associated with vertical migration of the gas through the cap rock, which provides a satisfactory seal since the formation originally contained gas for geological time periods.

During the gas injection phase, the maximum allowed pressure of the reservoir must not be exceeded, thus eliminating any risk of damage to the cap rock. Secondary risks, such as gas leaks through abandoned wells, are analysed during the conversion assessment phase. A thorough study is conducted on all the geological problems and the field production data. Care is also taken to ensure that all the wells are properly cemented.

A number of parameters must also be determined before the study is finalized:

- the optimal depletion rate of the field before it is used as a storage facility,

- the filling dynamics,
- the problem of gases mixing, and
- secondary recovery of the oil in place, in the case of an oil field.

Converting a producing field into a storage reservoir, therefore, requires thorough knowledge of its geological and physical characteristics, particularly the recovery ratios. There is no guarantee that the storage reservoir will have the same capacity as the initial field. This information is not always provided during the production phase of the field. Three-dimensional models and technological improvements have been developed that now give a good picture of the subsoil and enhance efficiency considerably.

A great advantage of depleted gas reservoirs is that at least the irreducible gas saturation is already existing, in some cases even all or a part of the future cushion gas. In addition, the wells are already in place and their equipment can sometimes be reused.

11.4.6 Aquifers

The principle of aquifer storage is to create an artificial reservoir by injecting gas into the pore spaces and/or the fracture system of the reservoir rock. To achieve this, the following geological conditions must be satisfied:

- a geological trap with a sufficient closure
- a porous and permeable reservoir
- a gas tight cap rock

The services of geologists, geophysicists, and reservoir engineers are enlisted to find such structures. Additionally it is important to know, why this aquifer was originally water-bearing and not filled with hydrocarbons. The reasons may be:

- immature or missing source rocks,
- no closed structure,
- no sealing caprock, and
- very recent trap development in geological times compared to gas migration.

Only for the first and the last possibility the further aquifer exploration can be continued with the chance of positive results. The geometric characteristics help to predict the storage volume, provided the operating pressure and the effective porosity to the gas are known.

After a fairly long filling period, alternate injection and withdrawal cycles take place on the reservoir. During the injection phase, gas is pressed into the reservoir. The pressure rises progressively and the water is expelled. Reservoirs in a virgin aquifer are operated at pressure 1.45 to 1.5 times the hydrostatic pressure. During the withdrawal phase, the gas is recovered at the wellhead and sent (with or without compression) to the pipeline network. As withdrawal proceeds, the pressure drops, the water moves back and again occupies the pores of the reservoir rock.

When the stored gas is withdrawn, the gas column within the storage horizon decreases until the time when the supply of the desired rate becomes uncertain in normal operation, because the water is liable to invade the base of the wells. The quantity of gas in place in the reservoir at this time represents the cushion gas.

In an aquifer reservoir, it is important to examine the effects on the surrounding aquifers and to check the tightness of the reservoir.

11.4.7 Gas Storage in Caverns or Abandoned Mines

Gas is stored in artificial hollow spaces than in the connected pore space of natural rocks.

11.4.8 Salt Caverns

This type of facility has been used to store LPG for many years, but the technique is relatively recent for natural gas.

The principle consists in dissolving the salt with fresh water and removing the brine by a single well, which then serves for gas injection and withdrawal (Fig. 11.7). These reservoirs serve to store relatively smaller quantities of gas than those that can be stored in aquifers or depleted reservoirs. The storage capacity for a given cavern volume (generally about 300,000 m³ geometrical volume) is proportional to the maximum operating pressure, which depends on the depth.

These facilities are operated by compression/expansion, between maximum gas volume, when the accumulated gas is raised to the maximum permissible pressure, and low gas volume when, after withdrawal by expansion of the stored gas, the pressure reaches the minimum permissible level (Fig. 11.8). Hence, the cushion gas depends on the extreme operating pressures adopted.

One of the problems encountered in this type of storage facility is the formation of hydrates. This is difficult to predict. The injected gas is relatively dry and gradually accumulates moisture. Moreover, it is not certain that the humidity results exclusively from the brine left on the bottom, but also from the salt, which is not completely anhydrous. As a rule, hydrates are not formed once the well is flowing at a substantial rate. The problems of hydrates encountered only occur at the time of experimental measurements and when withdrawal is initiated.

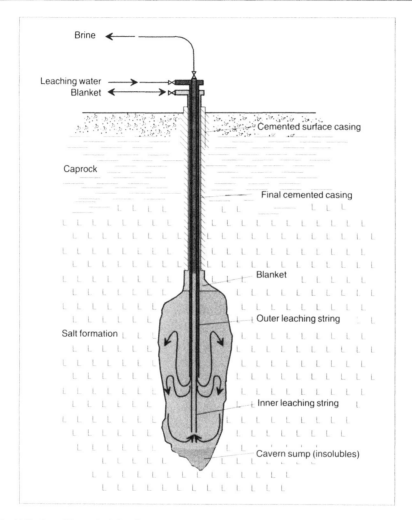

Fig 11.7 Leaching principle of a salt cavern

Some geological conditions for the construction of a gas cavern in rock salt must be fulfilled and proved by exploration methods (seismic, drilling works, laboratory investigations on drilling cores):

- Subsurface salt layer, salt pillow, salt dome, or salt wall of sufficient thickness and areal size and in a depth range, suited for salt caverns (the deeper the cavern the higher the maximum pressure and also the rate of convergence).
- Rock salt of relatively high purity and as small as possible with intercalations of thick insoluble formations (anhydrites, clay) or of

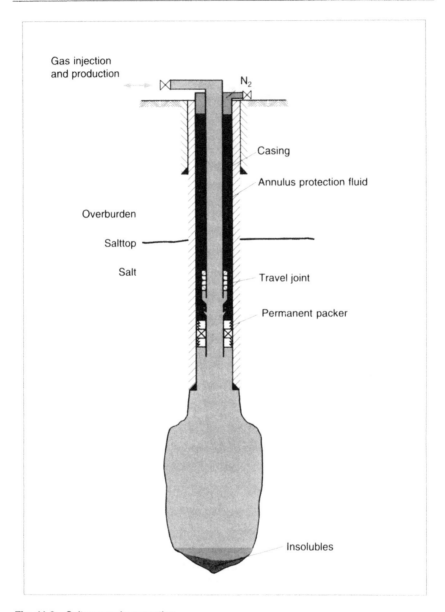

Fig. 11.8 Salt cavern in operation

extreme highly solube salt minerals (sylvite, carnalite), especially
when these intercalations are layered very complex with high dip
angles. The bedding planes of these intercalations may be
permeable for gas.

- Sufficient rock mechanical properties of the rock salt which is important for the future stability of the cavern (will be proved by rock mechanical investigations on salt cores and by rock mechanical calculation, resulting in cavern dimensioning.

Salt caverns are not merely a useful complement to the large porous reservoirs. In fact, they offer several advantages:

- high production rate,
- high degree of availability,
- short filling period,
- high level of safety,
- low percentage of cushion gas, and
- total recovery of cushion gas (in case of flooding the cavern with brine).

Thus, the combination of the two types of storage, in porous reservoirs, which are generally used to guarantee basic demand to balance seasonal variations, and storage in salt cavities, which are generally operated to cover peak demand, allows for high withdrawal rates even at the end of the withdrawal period.

11.4.9 Caverns in Non-Saliferous Rocks

Gas storage in caverns in non-saliferous rocks (like chalk, soft limestone, gypsum, marl, soft sandstone, and soft clay) are without any greater importance. They are not presented and discussed here.

11.4.10 Disused Mine Cavities

Very few storage facilities are located in abandoned mines. Three disused coal mines are in service, one in the United States (Leyden near Denver, Colorado), and two in Belgium (Anderlues and Péronnes). A disused potassium mine is employed in East Germany (Burggraf-Bernsdorf) to store natural gas.

Disused coal mines are used in Belgium to store natural gas at relatively low pressure. They offer a number of advantages:

- The existence of voids, which hence do not need to be created as in salt cavities.
- A real storage capacity that is higher than that of the existing voids, caused by the absorption of the gas by the surrounding coal.

To create a storage facility in a disused coal mine, it is primarily necessary to solve the problems of stabilization and tightness of the mineshafts plugged when mining operations were suspended. The second problem to be solved is that of the gas injection and withdrawal system.

Access to the reservoir is generally achieved through a shaft that remained accessible and which served to withdrawal mine gas. It is also necessary to take non-negligible mine problems into account, such as the ventilation of the mine gas that the deposit continues to release.

During withdrawal, the quality of the gas is altered and its heating value is lower than that of the gas that has been injected (reduced up to 6 % in the Anderlues storage facility). This decrease is due to the fact that coal exhibits a property to retain the heavy hydrocarbon fractions (pentane, propane, butane, and ethane). To restore the original heating value of gas, propane is injected on withdrawal. As the reservoir is drained, the heating value rises, and ultimately exceeds that of the original gas.

11.5 WORLDWIDE UNDERGROUND GAS STORAGE CAPACITY

The development of underground gas storage is closely linked to the development of gas supply around the world. The first underground storage capacities were developed as follows:

- In partly depleted gas fields - USA, 1916
  - Europe, Poland 1954
- In aquifers   - USA, 1947
  - Europe, Germany 1953
- In salt caverns   - USA, 1961
  - Europe, France and Germany 1970
- In disused mines   - Europe, Belgium (coal mines)
  Germany (potash mines)
- In excavated rock mines   - Europe, Czech Republic 1997

According to an IGU document published in 2000, a total of 600 storage facilities are in service around the world, 541 of which are porous rock storage facilities and 59 are cavern storage facilities (1998 data). The geographic locations and the most important storage data are summarized in table 11.1.

11.5.1 North America (US and Canada)

North America has the world's biggest natural gas market and a large number of underground storage facilities. The 448 facilities store 120 BCM of working gas or 39 % of the world's total working gas volume.

The facilities in the United States are located in more than 30 different states, with the Midwest showing the largest density of storage facilities (30 %), followed by the Southwest (28 %). 86 % of the gas is stored in depleted fields, 10 % in aquifers, and only 4 % (for geological reasons) in

Table 11.1 *Worldwide underground storage facilities in service (1998)*

	Number of storage facilities	of which porous-rock storage facilities	of which cavern storage facilities	Working gas (BCM)	Withdrawal capacity (BCM/day)
Western Europe	78	56	22	55	1.09
Eastern Europe & Central Asia	67	64	3	131	1.00
USA	410	383	27	106	2.15
Canada	38	31	7	14	0.20
Australia & Japan	7	7	0	2	0.01
Total	**600**	**541**	**59**	**308**	**4.45**

salt caverns. Because of the large number of market players both on the supply side and on the demand side, a strong competition has developed despite state regulation. Gas trading is based on hubs distributed throughout the country. The increasingly important role of underground storage facilities in the US gas industry is reflected by the huge investments in some 100 new storage projects currently underway.

In Canada, five provinces have underground storage capacity, 82 % of which has been developed in depleted gas fields, while the remaining 18 % are in salt caverns.

11.5.2 Western Europe

In Western Europe, the largest storage capacities can be found in Germany, Italy, and France. The key data for the facilities in these countries and in the rest of Western Europe are shown in Table 11.2.

New storage construction or expansion projects are currently underway in most Western European countries. The development of additional capacity is determined by a number of factors which include an increase in storage volume and capacity as a result of competition, deregulation, and third-party access on the one hand, and more stringent and environmental requirements on the other. Until 2010, a total of 21 storage projects and expansions are scheduled to start, 14 of which in Germany alone.

11.5.3 Eastern Europe and Central Asia

Eastern Europe and the countries of the former Soviet Union also have a number of storage facilities, some of which were developed as long ago as 40 years. Because of Russia's role as a gas exporting nation and as large consumer of natural gas, the country accounts for a large part of the

Table 11.2 *European underground storage facilities in service (1998)*

	Number of storage facilities	of which porous-rock storage facilities	of which cavern storage facilities	Working gas (BCM)	Withdrawal capacity (MCM/day)
Germany	38	23	15	16	390
Italy	9	9	–	15	263
France	15	12	3	11	190
Others	16	9	7	13	247
Total	**78**	**53**	**25**	**55**	**1,090**

world's storage capacity. The most important storage data for Eastern Europe and Central Asia are summarized in a table 11.3.

Further investments planned for the period until 2010 will help boost storage volumes but are mainly intended to increase withdrawal capacity. Most of the additional capacity is planned in Poland, Romania, the former Yugoslavia, Turkey, and the Baltic states, particularly Latvia and Lithuania.

Table 11.3 *Eastern and Central Asia underground storage facilities in service (1998)*

	Number of storage facilities	of which porous-rock storage facilities	of which cavern storage facilities	Working gas (BCM)	Withdrawal capacity (MCM/day)
Former Soviet Union	42	41	1	120	846
Bulgaria	1	1	-	0.5	4
Croatia	1	1	-	0.5	4
Czech Republic	4	3	1	3	40
Hungary	4	4	-	3	34
Poland	7	6	1	1	37
Romania	4	4	-	1	7
Slovak Republic	1	1	-	2	28
Total	**64**	**61**	**3**	**131**	**1,000**

11.5.4 Australia and Asia

In 1998, Australia and Asia had a total of seven underground storage facilities with a working gas volume of 2 billion m³. In **Australia,** a partly depleted gas field was converted into storage facility in 1981 to supply the large industrial centres in the southeast of the country. **Japan's** five small porous rock storage facilities developed in depleted natural gas fields only have regional significance. The country is known for its use of LNG terminals to cover peak demand. In the northwest of **China,** a rock cavern is used for gas storage. In Australia, China, Taiwan, and India, various locations are being examined for their suitability as storage facilities. In Iran, detailed surveying and exploratory drilling is currently underway at two potential storage sites.

11.5.5 Africa and South America

Africa's only storage facility is in **Morocco,** where several salt caverns are used to store LPG. Nothing is known about the plans to build new storage facilities on the African continent.

South America has no underground storage facilities at present. However, investigations are reported to be underway in **Argentina** and **Uruguay** to find suitable locations for storage sites. Increasing industrialization and the rise in population are the main driving factors for efforts to optimize the balancing of gas supply and consumption.

12

Underground Storage of
Nuclear Waste

Jan Richard Weber and Hans-Joachim Alheid

ABSTRACT

Radioactive waste arises from different sources. The production of electricity in nuclear power reactors is associated with the generation of high level radioactive waste. Low level and intermediate level waste comes from the use of radioactive isotopes in medical, research, and industrial processes. Repositories for low and intermediate level waste are operational in many countries. Most of them are near surface underground repositories. For the disposal of high level waste, only repositories in deep geologic formations seem to be eligible. Advanced programmes for the disposal of high level nuclear waste have been developed in Finland, Sweden, the USA and some other countries. As potential host rocks for high level waste repositories low permeable rocks like salt, clay, and crystalline rocks are considered. To evaluate, whether a formation is suitable to host a repository, the long-term behaviour of the system, composed of the host rock, the waste, and the backfill material must be known. For this purpose, underground research laboratories (URL) have been constructed in a number of countries. At the URLs, various experiments and investigations are carried out to study the processes involved in radionuclide migration, to test waste handling techniques and to develop and improve site characterization methods and tools. Another important purpose of

> URLs is to demonstrate the feasibility of geologic disposal of nuclear waste and thus to create confidence in geologic disposal.

12.1 INTRODUCTION

At present, there are 438 nuclear power reactors in 31 countries under operation. They produce some 16 % of the world's electricity. The production of electricity in a nuclear power reactor is associated with the generation of high level radioactive waste. Due to the hazardousness of nuclear radiation it is necessary to keep the nuclear waste away from the biospere. To achieve this, many countries are planning to dispose off high level nuclear waste in deep geologic formations. Until now, no repository for final disposal of high-level nuclear waste has been constructed. Many countries are just developing disposal strategies. Issues linked to the specific requirements of a repository for nuclear waste are investigated in a number of underground research laboratories.

This chapter gives an overview of the state-of-the-art of nuclear waste disposal in deep geologic repositories, with emphasis on high level waste, and of the associated research in underground laboratories. Overviews on this matter are given by OECD (1999), OECD (2001a), and OECD (2001b).

12.2 NUCLEAR WASTE

12.2.1 Origin

Radioactive waste is classified as low, intermediate, and high level waste, according to the intensity of the emitted radiation. High level nuclear waste comes predominantly from nuclear power reactors. The first nuclear power reactors were put into operation between 1950 and 1960 in the USA, Russia, and the United Kingdom. Today, there are 438 nuclear power reactors in 31 countries connected to the grid. In the year 2000, they produced nearly 2500 TWh electricity, which is equal to 16 % of the world's electricity.

To generate 1000 kWh electricity some 3 g of nuclear fuel is consumed. About 20 t of spent fuel accumulate annually per reactor at an average. The worldwide output of spent fuel adds up to about 7000 t per year. Spent fuel can be disposed of as high level waste or reprocessed. During reprocessing some nuclides are utilized for further use in nuclear power reactors. The remaining part of the spent fuel has to be disposed of as high level waste anyhow. Therefore, the final disposal of high level waste is part of the reprocessing option as well.

In addition, a nuclear power reactor produces some 10 m³ of operating waste. Operating waste consists of protective clothing, tools etc., which is low level waste, and of the ion-exchange resin used to purify the process water, which is intermediate level waste. Finally, a certain quantity of each kind of radioactive waste will be left from decommissioning a reactor at the end of its lifetime.

Other sources of radioactive waste are the production of radioactive isotopes for use in medical, research, and industrial processes. Radioactive substances are used in medicine in diagnosis and therapy. An important application of radioactive substances is non-destructive measurement of the thickness and density of many materials, as well as inspecting components for weaknesses and flaws. Radioactive materials are also used to sterilize medical products and to improve the durability of food by irradiation.

12.2.2   Properties of Nuclear Waste

Radioactive waste contains a variety of different unstable chemical elements, which decay spontaneously and thereby emmit gamma radiation, neutron radiation, and/or alpha radiation. This ionizing radiation can have harmful effects on living organisms. Therefore, it has to be isolated from the environment.

Due to the decay, the radioactivity decreases. The quantity of radioactive nuclides is reduced by one half within the specific half-live period, which varies for the different radionuclides between seconds and thousands or even billions of years. The major part of the radioactivity from spent fuel results from the actinides, which are the products from the nuclear fission in the reactor. Within less than 500 years over 99.9 % of the actinides decay. The radio toxicity of the entire spent fuel decreases to less than 0.01 % within one million years. All kinds of radioactive waste can produce gas, either by microbacterial degradation or by corrosion, provided that the waste get in contact with water.

12.3   NUCLEAR WASTE MANAGEMENT CONCEPTS

Disposal in near surface repositories or in above surface facilities is in many countries considered to be a safe and responsible way to manage low level radioactive waste. By combustion, evaporation, decontamination and compression, the volume of the low level waste is reduced before it is disposed off.

For the final disposal of high level waste only repositories in deep geologic formations seem to be eligible at present. Other options for the final disposition of high level waste, such as sinking into Arctic ice or firing into space, were condemned after careful consideration. More ambiguous is the situation with long-lived intermediate level radioactive waste. Countries, which aim to establish a repository for high level waste in a deep geologic formation, can use it for long-lived intermediate level waste as well. Countries, which don't have nuclear power plants and therefore don't have high level waste, can either dispose off the intermediate level waste together with the low level waste, or establish a separate repository for the intermediate level waste, or search for other solutions. Australia, for example, has about 500 m³ of long-lived intermediate level radioactive waste, which will be extended by a few cubic metres per year. In Australia, this type of waste is considered to be not suitable for near-surface disposal. But on the other hand, the cost of constructing a deep disposal facility does not appear to be justified given the small quantity of waste involved. Australia has postponed the decision, how to manage intermediate level waste, into the future.

12.3.1 Low Level and Intermediate Level Waste

Repositories for low level and intermediate level waste are operational in many countries. Most of them are near surface underground repositories, but above ground disposal is as well practised as deep underground disposal.

Germany has a long experience in deep underground disposal of low level waste. In 1967, disposal of low level waste began in the abandoned Asse salt mine for demonstration purposes. At present, the filling of the mine with crushed salt is under way. The former German Democratic Republic started disposal of low level waste in the Morsleben salt mine in 1981. Disposal was stopped in 1998. Until that time, about 40,000 m³ had been disposed. Now the repository is to be shut down. For stabilization of the mine it is planned, to pump more than 600,000 m³ of a concrete-salt mixture into the galleries and rooms. A new repository might be constructed in the abandoned iron ore mine Konrad. The ore formation in a depth between 800 m and 1,300 m is well sealed from fresh ground water by clay layers. The application for installation a repository at the Konrad mine had been considered for about 20 years when it was approved in 2003. Since then the construction of the repository is prevented by way of action.

Sweden has been operating an underground repository for low level waste since 1988. The final repository for radioactive operational waste (SFR) was built for the processing waste of the Forsmark Nuclear Power Plant and is directly accessible from the site of the power plant. The disposal rooms were excavated in a granite formation 50 m below the Baltic Sea. Various disposal techniques are used at the SFR. Portions of the waste are emplaced into rock faults without additional technical barriers, other portions are surrounded by clay and concrete.

Finland has also built final repositories for operational waste from nuclear power plants inside the power plant area. Such repositories were licensed in 1992 at the Olkilouto Power Plant and in 1998 at the Loviisa Power Plant. The disposal rooms have been excavated in a depth of 70–100 m. The Loviisa repository is scheduled to be decommissioned by sealing the tunnels and shaft in 2055.

The repository at Drigg near the UK reprocessing plant Sellafield has been under operation for more than 40 years. During this time, the repository has received nearly one million cubic metre low level waste. The total capacity amounts almost twice as much, which is sufficient for all low level waste from the UK until the middle of the century. At Drigg, the waste is packed in steel containers and disposed off in concrete lined shallow vaults.

France began to deposit low and intermediate level waste in a surface repository in 1969 at the Manche Disposal Facility. The Manche Disposal Facility received waste until 1994. The waste is covered with multiple layers of bitumen, soil, and sand. The facility is now under surveillance. Since 1992, low and intermediate level waste is disposed of at the Centre de l'Aube Facility. The capacity of this facility amounts to 1 million m^3 and is expected to enable a 60 year operating period.

Norway has been operating a repository for low and intermediate level waste since 1998 at Himdalen, near Oslo. The repository consists of four caverns, excavated in crystalline bedrock in a depth of 50 m. In the caverns, the waste drums will be embedded in concrete. The capacity of 10,000 drums is sufficient for operation until 2030. After closure of the repository a surveillance period of 300–500 years with monitoring the release of nuclides from the caverns is projected.

Japan has been operating the Rokkasho-mura facility for shallow disposal of low level waste since 1992. The underground of the Rokkasho-mura facility consists of tertiary and quaternary sediments. The waste is encapsulated in steel drums and emplaced in conrete lined cavities. The repository is designed for a final capacity of 600.000 m^3.

Shallow disposal of low level waste in concrete-lined structures has been applied also in Spain since 1992 at the El Cabril disposal facility.

Switzerland has decided, to create a repository for low and intermediate level waste in a deep geologic formation. A repository with concrete-lined disposal rooms is projected at Wellenberg in a low permeable marl formation. Due to public resistance, the initial conception of a maintenance-free final repository was modified into geologic long term storage. The new concept includes a surveillance phase of several decades after closure of the repository and during this phase accessibility of a minor part of the repository.

The world's first deep geologic repository for long-lived transuranic waste is the Waste Isolation Pilot Plant (WIPP) in New Mexico, USA. It started operation in 1999. The WIPP site is licensed for disposal of low level and intermediate level waste including long-lived waste from military purposes. High level waste is excluded. The repository is constructed in a 600 m thick bedded salt formation. Disposal rooms were excavated in a depth of 700 m. The formation is expected to remain stable for at least 250,000 years.

12.3.2 High Level Waste

All management concepts for high level waste include their final disposal in deep geologic structures. The feasibility of geologic disposal has been demonstrated by natural analogues. For example, the uranium deposit Cigar Lake in Canada gives an illustration of a natural analogue. The ore of Cigar Lake contents about 50 % of uranium dioxide. The deposit in a depth of several 100 m is surrounded by clay, which has prevented spreading of the uranium for more than one billion years.

Advanced programmes for the disposal of high level nuclear waste have been developed in USA, Finland, Sweden, some other countries.

Sweden

In Sweden, the nuclear power industry is responsible for managing high level radioacitve waste. Therefore, operators of Swedish Nuclear Power Plants founded the Swedish Nuclear Fuel and Waste Management Company SKB. SKB carried out studies on final disposal of spent fuel and established the so-called KBS-3 concept.

Sweden plans to built a repository for high level waste in crystalline bedrock at a depth of more than 400 m. The KBS-3 concept comprises three elements for the retention of nuclides from the biosphere: Isolation, retention, and dilution. For isolation, the spent fuel will be packed in leak-proof copper canisters. This material is chosen to prevent leakage of the canisters due to corrosion. Retention is reached by embedding the canisters in bentonite, which is a sort of clay with high potential for adsorption of nuclides. Another favourable property is the swelling of the bentonite when it gets in contact with water. Swelling of the bentonite package will form a tight enclosure of the copper canisters. Moreover, the plasticity of the bentonite gives a mechanical protection of the canisters in case of displacements in the surrounding bedrock. Dilution is achieved by a proper selection of the disposal site with long pathways between repository and biosphere.

After removal from the reactor, the spent fuel is cooled down in water filled basins for 30 years. The remaining heat production sets certain requirements in the layout of the repository, to ensure, that the host rock won't sustain damage due to the heat production of the spent fuel. The copper canisters will be placed in vertical boreholes with a diameter of nearly 2 m and a depth of 8 m. The boreholes will be drilled into the floor of the disposal tunnels with a distance of 6 m in between. Disposal tunnels will have a spacing of 40 m. To take in 4,500 canisters, each of them loaded with 2 t of spent fuel, the repository will have an extension of up to 4 km².

The construction of the main part of the repository will be preceded by a test phase. During the test phase, a pilot repository, which comprises a partition of about 10 % of the final extend of the repository, will be built. After charging this partition, the behaviour of the repository will be monitored for several years, before construction of the main part of the repository will start.

Preliminary investigations were carried out at six potential repository sites, out of which SKB proposed in the year 2000 three sites for further investigations. In 2001, the municipality Östhammar, which is one of the three concerned municipalities, agreed to further investigations in the area of the Forsmark Nuclear Power Plant. In March 2002, the municipality council of Oskarham agreed to investigations at the Simpevarp site. The approval of these two municipalities on further investigations meets the directive of the Swedish government, that at least two different sites have to be evaluated concerning their suitability to host a deep repository for spent nuclear fuel. After realization of drilling programmes and analysis of the results, a final decision on the siting of the repository is expected around 2009. According to SKB's schedule, the pilot repository might commence operations in 2015.

Finland

Details of the Finnish concept for final disposal of spent nuclear fuel were developed in cooperation with Sweden. Therefore, some features of the Finnish concept are similar to the Swedish KBS-3 concept. As in Sweden, the Finnish repository will be constructed in crystalline bedrock. The spent fuel will be encapsulated in double wall containers. The inner part is made of iron and designed to resist mechanical forces from displacements within the bedrock. It is enclosed by an external copper canister. The copper canister with a wall thickness of 5 cm will loose its tightness due to corrosion only after several 100,000 years. The canisters will be embedded in bentonite. The repository will be built in a depth of about 400–700 m.

The two Finnish Nuclear Power Plants in Olkiluoto and Eurajoki will produce 2,600 t of spent fuel during their 40 years of operation. To dispose of this quantity of spent fuel, disposal tunnels with a total length of 15 km will be mined.

The repository will not require any maintenance after closure. However, the Finnish government decided, that retrieval of the waste must be possible after closure of the repository. This is not for safety reasons, but for ethical reasons. At present, final disposal of high level radioactive waste in deep geologic structures is considered to be reasonable, safe, and associated with a smaller risk than long-term surface storage. In future, other disposal methods or waste utilization techniques might become available. Therefore, the option to reverse the final disposition of the spent nuclear fuel should be provided to future generations. Retrieval of the waste canisters from the repository is possible using standard mining techniques. It can be expected, that application of standard mining techniques is not possible in secrecy, so that the disposed waste is protected against misuse after unnoticed retrieval.

Spent nuclear fuel from the Finnish reactors was originally sent to Russia for reprocessing. Finland ceased shipment to Russia 1996. In May 2001, the Finnish parliament ratified a decision in principle on the disposal of spent fuel in a deep geologic repository in Finland. Out of four localities, on which suitability investigations were done, Eurajoki, near the Olkiluoto Nuclear Power Plant, was chosen as site for the repository. For characterization of the host rock, the underground facility Onkalo will be built as of 2004. According to plans, construction of the disposal facility will start in 2010. The repository might become operational in 2020.

USA

In 1982, US Congress established the national policy, which describes the procedure for approving a nuclear waste repository for high level waste. It made the US Department of Energy (DOE) responsible for building and operating the repository. To find a suitable site, DOE started a preliminary study in 1983 at nine locations, from which in 1985 three were selected for intensive scientific site characterization. Since 1987, site characterization was continued at only one site, Yucca Mountain in Nevada.

The host rock at Yucca Mountain is a tuff formation. The repository is to be built between 200 and 400 m below surface and above the water table, which lies in a depth of about 500m below surface. The area has a very dry climate, with less than 15 cm rainfall a year. Due to the lack of water, it is expected, that corrosion of the waste canisters should be minimized.

As a result of the study, DOE stated in 2002, that Yucca Mountain is scientifically and technically suitable for development of a long-term geologic repository.

This statement was based on a safety assessment for the next 10,000 years. In 2004 a U.S. court ruled, that this period is too short. Therefore a new safety assessment, cosidering much more than 10,000 years, is required for the continuation of the approval process.

Germany

In Germany, salt has been considered as a suitable host rock for final disposal of radioactive waste for about 40 years. Searching for a salt dome to host a repository for high level waste began in 1976. In 1979, investigation of the Gorleben salt dome started. An exploration programme from the surface, which comprised seismological surveys, geophysical investigations, and drilling of deep boreholes, yielded comprehensive information on the structure of the cover rock over the salt dome and the hydrogeological situation. To harm the salt structure as little as possible, only four boreholes were lowered into the margins of the salt dome. Another two boreholes were drilled at the selected locations for the two shafts, which were built as of 1986. About 10 years later, the shafts reached their final depths of 840 m and 933 m respectively. To explore the internal structure of the salt dome, drifts with an overall length of several kilometres were excavated. Exploration of the salt structure and investigation of the properties of the salt yielded no indication, that the site would not be qualified to host a repository for high active nuclear waste. Nevertheless, German government decided to suspend investigations at the Gorleben site. A working group was established, to re-evaluate the German disposal concept and to set up a site selection procedure with public participation, including different types of host rocks. A repository should be operational in 2030.

Other countries

Most countries favour the deep geologic disposition option for high level radioactive waste. Elaboration of the national concepts for deep geologic repositories have reached different stages and the scheduled dates for beginning operation of the repositories differ considerably. Spain has developed generic repository designs for the different host rocks salt, granite, and clay. A decision on the final disposal of high level waste shall be reached not before 2010. United Kingdom has also opted for geologic disposal of high level waste, but there is no schedule for an approach to constructing a repository. France has decided, to study the suitability of two different host rocks, clay, and granite. As investigation of a clay site is under way, a granite site for investigations is still to be selected. Japan aims at constructing a repository as of 2030 in granite or sedimentary rock. Belgium has already developed storage casks with stainless steel casings, but disposal of high level waste in a deep geologic repository is projected to start not before 2035-2080. China intends to put a deep repository into operation in 2040 or later. Meanwhile, a granite site in the Gobi desert is investigated. An underground research laboratory in the Gobi desert is projected to become operational in 2030. Russia is considering several sites as candidates for further studies, including salt, clay, granite, basalt, and permafrost ground. A schedule for implementing a deep repository has to be established. Switzerland considers granite and opalinus clay as potential host rocks for a repository, which will become operational not before 2020.

Some countries consider additional options for the middle-term disposition of high level radioactive waste. Canada for example has stated, that the disposition of high level waste in a 500–1000 m deep repository within plutonic rocks of the Canadian shield is a safe storage option. Feasibility of long-term storage at the reactor sites or in central surface storage facilities is considered yet.

Some countries have no long-term plans for the disposition of high level radioactive waste. Armenia for example, which operates one nuclear power reactor, considers spent fuel as a product that might be usable in future. Therefore, no programmes for the installation of a final repository are presently executed.

12.4 INVESTIGATION OF PROPERTIES OF DIFFERENT HOST ROCKS IN UNDERGROUND RESEARCH LABORATORIES

To evaluate, whether a rock formation is eligible to contain a repository, the long-term behaviour of the system composed of the host rock, the waste, and the backfill material, container material and other inserted substances must be known. For that purpose, underground research laboratories (URLs) have been constructed. At the URLs, various experiments and investigations are carried out to study the processes involved in radionuclide migration, to test waste handling techniques and to develop and improve site characterization methods and tools. Another important purpose of URLs is to demonstrate the feasibility of geologic disposal of nuclear waste and thus to create confidence in geologic disposal.

A number of URLs has been built at sites, where no construction of a repository is planned. Such URLs are referred to as generic URLs. They may be developed as a stand alone purpose-built construction or by utilizing existing excavations such as abandoned mines or highway tunnels. Examples for purpose-built generic URLs are the Hades Underground Research Facility in Belgium, which is constructed in plastic clay at a depth of 230 m, the Whiteshell URL in Canada, which is constructed in Granite at a depth of 240–420 m and the Swedish Äspö Hard Rock Laboratory, also constructed in granite at a depth of up to 450m. Generic URLs, that were established in existing underground facilities, are the German Asse mine, an abandoned potash and rock salt mine with excavations up to 800 m depth, the Stripa Mine in Sweden, constructed in granite in a depth of about 400 m, the Grimsel test site in

Switzerland, constructed in granite with 450 m overlying rock, the Mont Terri Project in Switzerland in hard clay in a depth of 400 m and the French Tournemire facility in hard clay in a depth of 250 m.

To a lesser extent, URLs were established at sites, where later on a repository might be built. The Waste Isolation Pilot Plant in New Mexico was such a site-specific URL, prior licensing as a repository in 1999. A French site-specific URL is being built at Meuse/Haute Marne in clay at a depth of 500 m. Finland has decided to develop the site-specific URL Onkalo in a granite formation at a depth of 500 m.

12.4.1 Salt

Salt features some favourable characteristics with regard to geologic disposal of nucler waste. The most important pathway, on which radionuclides might leave a repository and get access to the biosphere, is the flow of groundwater. In salt domes, no groundwater fluxes exist. Another favourable property is the plasticity of rock salt. If the initially isostatic stress field in a salt deposit is disturbed by constructing excavations, the deviatoric stress drives the salt to creep into the opening. Therefore, in the long run, openings in a salt deposit will be closed and deposited containers will get tightly embedded. Fractures will be closed due to the self-healing capacity of rock salt likewise. This mechanism will progressively seal a repository in a salt deposit. Furthermore, salt is relatively easy to mine.

Asse Mine

The first generic URL was the Asse Salt Mine in Germany. Development of the mine in the Asse salt dome started in 1906 by lowering the first shaft, which reached its projected final depth of 765 m in 1908. During the first years, potash was mined, later on rock salt was also mined. Mining was stopped in 1964. Experiments related to disposal of radioactive waste in rock salt started in 1965. Vertical stacking, horizontal stacking, and toppling the canisters were applied as different deposition techniques. Between 1967 and 1978 about 125,000 low level waste containers were disposed of as an experiment. Additionally 1,300 containers with intermediate level waste were deposited between 1972 and 1978. The deposited containers will remain inside the mine.

After 1978, experiments focussed on development and assessment of deposition techniques of heat generating waste and on testing of waste handling methods and of backfilling techniques. Recent experiments deal with the evolution of the backfilling under conditions, prevailing in a repository for high level waste.

The Bambus Project (Backfill and Material Behaviour in Underground Salt Repositories) demonstrates the evolution of a deposition drift, which contains several containers with heat generating waste. The simulated waste containers are designed like Pollux-Containers with a length of 5.5 m, a diameter of 1.5, m and a mass of 65 t. They are electrically heated with a power of 6.4 kW. In two adjacent drifts with a distance of 10 m in between, three containers are installed respectively. The drifts, which have a diameter of 4.8 m, are backfilled with crushed salt. Heating started in 1990 and lasted nearly 10 years. During the heating phase, the increase of the stress within the backfilling, caused by the creeping of the surrounding salt rock, and the decrease of the porosity of the crushed salt was monitored and compared with the predicted evolution.

In a similar project, the behaviour of the backfilling in the annular gap between a borehole and a simulated die-cast with reprocessing waste is investigated. During the first 150 days of the experiment, the porosity of the crushed salt backfilling decreased from 40 % to 15 %. Simultaneously the permeability decreased by one order of magnitude.

Waste Isolation Pilot Plant

The American Waste Isolation Pilot Plant (WIPP) was built in the Salado Formation, a bedded salt formation. Construction began in 1981 with lowering an exploratory shaft to a depth of 700 m. In 1982, underground excavation at the WIPP site began. Rooms were excavated 655 m below surface and 400 m below the top of the salt formation. The disposal rooms are situated in the southern part of the facility. In the northern part, various experiments were carried out. The bedded salt layers at WIPP were found to be partially brine saturated with very low permeabilities. Due to corrosion and salt creeping, destruction of waste containers is expected within a few decades. Therefore, experimental studies included thorough investigation of the geohydraulic conditions, to prove, that the integrity of the repository will be guaranteed after corrosion of the engineered barriers. Further experiments aimed at temperature effects on brine migration and salt creeping, and at the evolution of engineered plugs of excavations.

12.4.2 Granite

A particular advantage of granite as host formation for nuclear waste repositories is the long-term stability of many granite formations. Contrary to salt, which tends to close openings and fractures by creeping and thereby creates a tight sealing of a repository, excavations in granite will remain open and fractures will remain permeable. Therefore, at a potential site for a repository in granite formations, emphasis has to be put on a thorough characterization of existing fracture networks and migration processes therein, as well as on the conception and execution of engineered barriers.

Äspö

The Äspö Hard Rock Laboratory is located in the vicinity of the Oskarshamn Nuclear Power Plant. Construction was completed in 1995. It consists of an inclined access tunnel approximately 3,600 m in length with a maximum depth of 460 m. Research areas are positioned at different locations and depths. The laboratory is additionally connected to the surface via a shaft for hoisting and ventilation. Since 1995 a programme was performed to develop, test, and demonstrate the technology to be used for the disposal of spent nuclear fuel. At the Äspö Hard Rock Laboratory, testing is carried out in a realistic environment prior to applying the technology in a deep repository (Hammarström and Olsson, 1996).

One important area for the research carried out at the Äspö Hard Rock Laboratory concentrates on the function of the host rock as a barrier. It is particularly important to study the movement and chemical composition of the groundwater.

Another area is the acquisition of more in-depth knowledge of the action of bentonite backfill and the copper canisters with the rock under realistic conditions.

These investigations provide a data base that makes it possible to calculate the working life of the canisters, how the fuel is affected, and how long it will take for any radioactive substances dissolved in the water to be passed out in it, as well as the rate at which this will take place.

The Prototype Repository Project is a recent international, EC-supported activity to investigate, in full scale, the integrated performance of engineered barriers and near-field rock of a simulated KBS-3 deep repository in crystalline rock with respect to heat evolution, mechanics, water permeation, ion diffusion, gas migration, and microbial processes under natural chemical conditions.

The experimental setting comprises two test sections with a total of 6 canister positions in vertical boreholes and backfilled access tunnel. The heat production of radioactive waste is simulated by electrical heaters. The two test sections are separated by a concrete plug and the whole project is separated from the remaining URL by a second plug.

The main objectives for the Prototype Repository Project are:

- To simulate part of a future KBS-3 deep repository with respect to geometry, materials, and rock environment.
- To test and demonstrate the integrated function of repository components and to compare results with predictive calculations based on conceptual and theoretical models.

- To develop, test, and demonstrate appropriate engineering standards and quality assurance methods.
- To simulate appropriate parts of the repository design and construction process.
- To operate the prototype repository for up to 20 years, with the possibility to run one section for about 5 years and the other for a longer period of time.

Grimsel test site

The rock laboratory Grimsel is located 1,730 m above sea level in the Aare-Granite of the Juchlistock in the Swiss Alps. The laboratory is accessible via a pre-existing horizontal tunnel to a subsurface power plant (Marschall et al, 1999). The Grimsel test site consists of an approximately 1 km long system of drifts and chambers. The overlying thickness is 450 m. The test site was designed exclusively for research purposes, a subsequent construction of a repository at the site is not intended. The main purpose of the experiments carried out at the site was to develop and improve measurement techniques and characterization methods for fractured crystalline rocks.

At the Grimsel test site, radar measurements are used to inspect the rock without injuring it with boreholes. The technique is suitable to detect impurities in the rock at a distance of up to 200 m. The application of different transmitter and responder configurations yields three-dimensional maps of the fracture arrangements. A similar technique, which uses acoustic waves instead of electromagnetic waves is applied in seismic tomography measurements. Smallest movements of the rock mass are determined by inclination-measurements carried out in 6 boreholes. In migration experiments, the movement of tracer fluids through the fracture network is investigated. The applicability of conventional borehole tests to determine the permeability of the rock is limited in crystalline rock. Therefore, a ventilation test was performed for evaluating the permeability of the granite at the Grimsel Test Site.

International organizations from Germany, the USA, France, Sweden, and Japan participate in the experiments at the rock laboratory.

12.4.3 Clay

In many countries, argillaceous formations are being considered as potential host rocks for repositories of radioactive waste. The argillaceous media is a potential geological barrier because of its very low permeability and strong capacity for radio nuclide retention. In addition long-term plastic behaviour is assumed even for hard clay. This may allow healing of damage introduced by excavation processes. However, the material behaviour is sensitive to the degree of saturation, to pore water chemistry, and possibly temperature. These factors may change during the construction and operation of a repository.

Hades

The underground laboratory HADES, at a dept of 225 m below the SCK•CEN site at Mol (Belgium), allows the study of the behaviour of clay, of waste containers, and of the migration of radionuclides under realistic conditions (De Bruyn, E and B. Neerdael, 1991). For the HADES project, a research laboratory has been excavated in the Boom clay layer underneath Mol. Up to now, it is the only underground laboratory in the world for the disposal of high level radioactive waste in plastic clay. Within the frame of the safety of disposal in the Boom clay, research is concentrated on the chemical reactions of the waste forms to be disposed of with the environment, and the migration of the radioelements. The Boom clay is 32 million years old and already by nature contains uranium, thorium, and other long-lived isotopes. Thus, this clay is also an example of a natural analogon. Consequently, the migration of those substances already present over the years are studied, and on the basis thereof, geochemical models are developed.

The behaviour of different container materials is studied. It is the intention that the packaging itself isolates the conditioned waste for at least 500 years from the Boom clay. It appears, that stainless steel is not subject to corrosion in a Boom clay environment. Research is now extended to the study of direct disposal of spent nuclear fuel.

HADES has been extended by the construction of a second access shaft and an additional gallery, within the frame of PRACLAY (Bernier et al., 2000). The PRACLAY project is a 1:1 demonstration, to investigate the behaviour of the galleries and of the clay at temperatures equal to those which will prevail upon the disposal of vitrified high level waste. The thermal characteristics of the waste are simulated by electrical heaters. Multiple barriers around the radioactive waste constitute a basic principle of disposal. Excavation techniques that minimize disturbance are tested and the emplacement of waste packages in the disposal gallery, backfill, and sealing of the gallery using remotely controlled tools is perdemonstrated. Above the ground, PRACLAY builds a large-scale model in which the disposal concept is realistically simulated.

Mont Terri

In 1995, several organizations decided to start an international research project in the Reconnaissance Gallery of the Mont Terri motorway tunnel, in north-western Switzerland, in a Mesozoic shale formation, the Opalinus Clay (Aalenian) with a thickness of 160 m and a present depth of 250 m (Thury and Bossart, 1999). This project is under the patronage of the Swiss

Geological Survey. The following organizations are partners with equal rights in the project: ANDRA and IPSN (France), BGR and GRS (Germany), ENRESA (Spain), Nagra (Switzerland), JNC, Obayashi, and CRIEPI (Japan), SCK•CEN (Belgium).

The aims of the project are to analyse the hydrogeological, geochemical, and rock mechanical properties of an argillaceous formation, the changes of these properties induced by the excavation of galleries and to evaluate and improve appropriate investigation techniques. In January, 1996, eight niches were excavated for the project and 15 experiments were started in about 50 boreholes of up to 30 metres in length. In the winter of 1997-1998 a new research gallery and several additional niches with a total length of 230 m were excavated to host further experiments.

At present a total of 18 experiments are carried out; three of them are co-financed by the European Commission and the Swiss Federal Office for Education and Science (BBW). On one hand, these experiments will contribute to answer questions related to the assessment of repository safety in a consolidated argillaceous formation and, on the other hand, they will contribute to evaluate adequate investigation and construction techniques. The key issues are:

- Evaluation of investigation techniques.
- Characterization of a clay formation
- Characterization of changes in the rock, induced by gallery excavations, heat of waste, and cement waters.
- Repository construction, waste emplacement, and gallery sealing.

Tournemire

The French Institute for Radioprotection and Nuclear Safety (IRSN) selected the Tournemire site (Aveyron, France) in order to develop research programs concerning the confining properties of indurated argillaceous formations. The Tournemire site is located in a Mesozoic marine basin on the Southern border of the French Central Massif. The sedimentary series at this site are characterized by three major layers of Jurassic age. The argillaceous media corresponds to a 250 m thick layer, which consists of claystones (argillites) and marls of the Toarcian and Domerian. It is located between two limestone aquifers: a regional aquifer in the lower part (Carixian aquifer) and an Aalenian aquifer in the upper part (Boisson et al., 1998; Cabrera et al., 1998).

The Tournemire site was selected because a railway tunnel (one century old) gives access to the argillaceous media. In 1996, two 30m long horizontal drifts were excavated orthogonally to the tunnel axis. The

Tournemire investigations are conducted by the IRSN research group with external scientific and technical collaborations.

The research programs are dedicated to the evaluation of fluid transfer processes through these formations, water-rock interaction processes, radio nuclide sorption or diffusion laboratory experiments, as well as the characterization and long-term monitoring of the excavation disturbed zone (EDZ).

12.5 CONCLUSIONS

The feasibility and safety of geologic disposal of radioactive waste has been demonstrated. For disposal of low level and intermediate level waste, a number of repositories are operational. Construction of repositories for high level waste is being prepared in a number of countries and seems to be imminent in Finland and the USA.

For further development of the nuclear waste disposal techniques, the international URLs are very important. Actual research concentrates on repository design and disposal technology and the analysis of nearfield processes from the waste up to the disturbed zone of the host rock. Key question for both topics is the optimization of engineered barrier systems with regard to long term stability, operational safety, and construction techniques. This requires a comprehensive characterization of the host rock as a basis for reliable conceptual and constitutive models that allow credible thermo-mechanically-hydraulically-chemically coupled models for the long-term safety approach in Performance Assessment.

REFERENCES

Bernier, F., Buzyenes M., Brosemer D., and De Bruyn D., 2000. Extension of an underground laboratory in a deep clay formation, GeoEng 2000, An international Conference on Geotechnical & Geological Engineering, Volume 2, Melbourne.

Boisson J.-Y., Cabrera, J., and De Windt, L., 1998. Etude des écoulements dans un massif argileux, laboratoire souterrain de Tournemire. EUR 18338FR.

Cabrera J., De Windt, L., Moreau-Le Golvan, Y. and Boisson, J.-Y. 1998. Characterization of fractures as Water-Conducting Features in Shales - Studies at the Tournemire Test Site. Third GEOTRAP Workshop.

De Bruyn, D. and Neerdael, B., 1991. The Hades Project – Ten years of civil engineering practice in a plastic clay formation, Proceedings Conference organised by the Institution of Civil Engineers ,Windermere, March,Thomas Telford Eds., 39-50

Hammarström, S., Olsson, O., 1996 (edition) ÄSPÖ Hard Rock Laboratory, 10 Years of Research. Swedish Nuclear Fuel and Waste Management Company, Stockholm

Marschall, P., Fein, E., Kull, H., Lanyon, W., Liedtke, L., Müller-Lyda, I., Shao, H. 1999. Grimsel Test Site, Investigation Phase V (1997 - 2002), Conclusions of the Tunnel Near-Field Programme (CTN). Nagra Technical Report 99-07, Wettingen (Switzerland)

OECD Nuclear Energy Agency. 1999. Progress Towards Geologic Disposal of Radioactive Waste: Where Do We Stand? An International Assessment. Paris.

OECD Nuclear Energy Agency. 2001a. Nuclear Waste Bulletin, Update on Waste Management Policies and Programmes, No. 14 – 2000 Edition. Paris.

OECD Nuclear Energy Agency. 2001b. The Role of Underground Laboratories in Nuclear Waste Disposal Programmes. Paris.

Thury, M. and Bossart, P., 1999 (edition). Mont Terri Rock Laboratory: Results of the Hydrogeological, Geochemical and Geotechnical Experiments Performed in 1996 and 1997, Geol. Rep. Swiss Natl. Geol. Surv. 23.

13

Sustainable Development of Groundwater from Hard Rock Formations

Gerd Klee and **Fritz Rummel**

ABSTRACT

Water as drinking water, mineral water, thermal water, for the use in agriculture and industry is of vital importance with ever increasing demand. In India nearly 800.000 villages do not have sufficient water supply, similar is valid for large areas in Africa, South America, and Asia. Even countries with sufficient rainfall like Germany will face drinking water shortage problems in the near future because of shallow aquifer pollution due to disposal of untreated industrial waste and the use of fertilizers and pesticides in agriculture during the past. As a result we are forced to tap deeper aquifers mostly in jointed hard rock formations. In spite of intensive geological and geophysical exploration deep wells often show poor water yield. We were told that only 50 per cent of drillings for water on the Indian shield are successful, and wells drilled just a short distance from a dry well may be productive if the borehole intersects a water-bearing joint or fracture network. Here, it seems obvious to use stimulation techniques to artificially connect the well with the water-bearing network in its vicinity. We describe some case histories for water-well stimulation in Germany.

13.1 INTRODUCTION

Groundwater is of vital importance in meeting todays demand of water supply for domestic, agricultura and industrial use. Due to the extensive use of fertilizers and pesticides in agriculture and the disposal of untreated industrial waste, shallow aquifers in porous and permeable sedimentary rock formations often are polluted. In addition, increasing water demand due to rapid population growth and unresponsible over exploitation may lead to seasonal depletion of such aquifers. Both these qualitative and quantitative problems are recognized by scientists and politicians all over the world. The problems were particularly addressed for India by the International Groundwater Conference on Sustainable Development of Groundwater Resources held at Dindigul, Tamil Nadu, India in early 2002 (Thangarajan et al., 2002). This conference covered topics like groundwater assessment, recharge processes, pollution remediation measures, and resource management. It also touched upon issues like the necessity to drill deeper wells into non-porous, often crystalline rocks with the hope of drawing water from open joint and fracture networks. However, such drillholes often are without sufficient yield in spite of accompanying intensive and competent geological and geophysical prospection. On the other hand, there are many cases where a borehole drilled next to a dry hole may be highly productive if the drillhole intersects water bearing fractures and joints which often are steeply inclined.

13.2 STIMULATION TECHNIQUES

Before abandoning a well with insufficient water yield, efforts are made to increase the productivity of the well by blasting or acidizing operations. In most cases, such operations do not lead to a full success since their effect is limited to the close vicinity around the wellbore. In contrast, the hydraulic fracturing method as used in oil industry for borehole productivity enhancement offers a tool to either initiate and propagate a fracture far away into the massive rock to intersect existing water bearing joints, or to stimulate and open closed fractures such that water can migrate towards the wellbore (Fig. 13.1).

The hydraulic fracturing concept is simple and straightforward (Rummel, 2002). A borehole interval is isolated from the rest of the borehole by two inflatable packer elements which allow the interval to be pressurized until a fracture is induced in the borehole wall rock. The induced fracture can be easily propagated over large distances by further fluid injection. The critical pressure for fracture initiation, P_c, is depending on the rock tensile strength P_{co} and the acting stresses within the rock

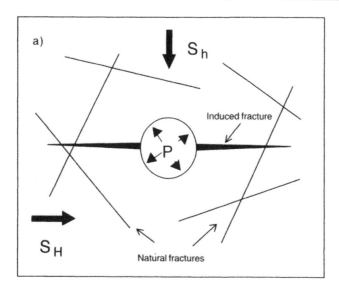

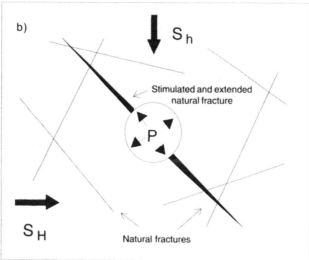

Fig. 13.1 The hydraulic fracturing borehole stimulation concept, showing (a) an induced fracture originating in a pressurized borehole and intersecting natural fractures. S_H, S_h principal horizontal stresses, P injection pressure, (b) a natural fracture being stimulated and extended.

mass. For a vertical borehole and the acting horizontal far-field stresses S_H and S_h ($S_H > S_h$, principal stresses) the condition for fracture initiation is given by the relation

$$P_c = 3S_h - S_H + P_{co} \qquad (1)$$

Fracture propagation can be described by a simplified fracture mechanics relation of the form

$$P_c = \frac{K_{IC}}{k_3} + k_1 S_H + k_2 Sh \tag{2}$$

where K_{IC} is the rock fracture toughness, a material property (e.g. $K_{IC} = 2$ $MN/m^{3/2}$ for granite), k_1 and k_2 are dimensionless stress intensity functions ($k_1 \rightarrow 0$, $k_2 \rightarrow 1$ for large fractures), and k_3 is a stress intensity function mainly depending on the pressure distribution within the pressurized and propagating fracture. For moderate fluid injection rates the first term in eq. (2) can be neglected for sufficiently large fractures (fracture length a >> borehole radius R). Thus, the pumping pressure P_p for fracture extension is practically only controlled by S_h or the normal stress S_n acting perpendicular to the fracture plane. When the pumping pressure is higher than the normal stress S_n the fracture is open and large amounts of water can be pumped through the fracture. To keep the fracture open after pressure release oil industry uses various kinds of proppant materials such as sand. During geothermal energy extraction research from hot crystalline rock it was recognized that during fracture stimulation frictional shearing on fractures favourably aligned with the principal stresses occurs, which prevents subsequent fracture closure. Thus, simple stimulation of fracture may lead to permanent permeability increase of the rock mass and productivity enhancement of initially dry water wells (Baria et al., 1999).

13.3   CASE HISTORIES OF WATER WELL STIMULATION

In the following sections some case histories are presented which may demonstrate the development of water well stimulation from an initially empirical approach towards a systematic procedure. All of these studies were conducted by MeSy in different rock formations in Germany.

13.3.1   Borehole Gunzenhausen, Bavaria

Borehole Gunzenhausen was planned for mineral and thermal water supply to stimulate recreation attraction in this area. The borehole was drilled to a depth of 460 m with a bottom hole temperature of 23°C. The borehole penetrates sedimentary rocks to a depth of 295 m, and then is in the crystalline basement consisting of a fine-grained granite. The hole was cased to a depth of 350 m. The diameter of the open-hole section below is 156 mm. The geophysical logs (caliper, salinity) indicated a major fault zone at 440 m which also caused borehole instability at this depth. The natural gamma log suggested clay fillings of some fractures. The water table was approximately 60 m below surface. Initial pumping tests yielded a productivity of 1.6×10^{-11} m^3/Pa s for the open-hole granitic section from 350 to 460m.

Stimulation tests in the open-hole section from 376 to 460 m were carried out with (i) a wireline double straddle packer system of 1 m test length and injection rates of some litres per minute, and (ii) with single packer and double straddle packer systems of 4 m test length which allowed injection rates of up to 3.2 litres per second. A typical test plot for test series (i) is given in Fig. 13.2 indicating the pressure decay after three pre-stimulation pressure pulse tests, and the pressure decay after system shut-in after the hydrofracturing stimulation. Evaluation of the test yields an initial rock permeability of (13 ± 3) µDarcy, and a permeability of (64 ± 13) µDarcy after stimulation. The normal stress acting across the induced fracture plane is app. 14 MPa. The permeability increase suggests that the fracture is not completely closed.

A total of 19 such stimulation test were carried out within the granitic open-hole section from 380 to 430 m. A final injection test with a single packer set at 376 m (Fig. 13.3) demonstrated a steady-state injection rate of

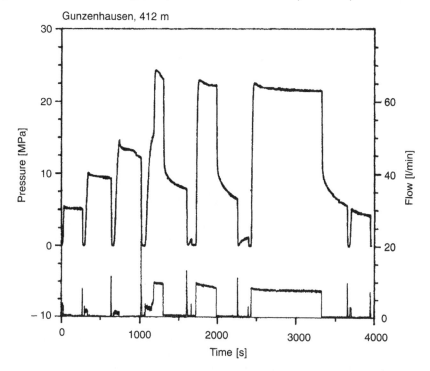

Fig. 13.2 Pressure and flow rate record of initial permeability tests. (i) hydrofrac test (ii) and Post-frac injection tests at a 4 m test section at 412 m depth in borehole Gunzenhausen.

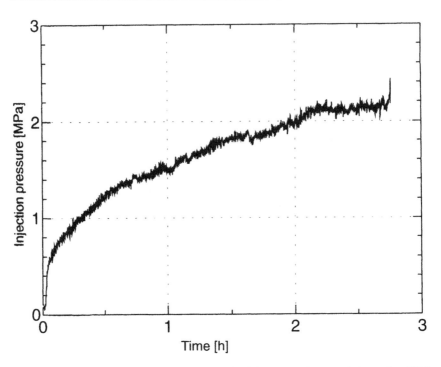

Fig. 13.3 Pressure response during the final injectivity test on the open-hole section 380 to 430 m in borehole Gunzenhausen (constant injection rate: 4.5 litres per minute).

4.5 litres per minute at a pressure of 2.2 MPa which yields an injectivity of 3.4×10^{-11} m^3/Pa for the total open-hole section, compared to the initial productivity of 1.6×10^{-11} m^3/Pa s an increase by a factor of two. This productivity increase, however, was not sufficient for an economical use of the crystalline aquifer at depth.

13.3.2 Borehole Waffenbrunn, NE Bavaria

The borehole with a depth of 150 m and a diameter of 250 mm was designed to contribute to the water supply of the community Waffenbrunn from the fractured gneissitic underground. The expected discharge was 3 litres per second, but only one fracture zone with a yield of 0.5 l/s at 55 m depth was penetrated. During an initial injection test with a single packer set at 41 m depth, the injectivity of this fracture could be increased by 30 per cent to 1.3×10^{-9} m^3/Pa s (Fig. 13.4).

Subsequently, 20 hydraulic fracturing tests with a double straddle-packer system with a test section length of 3 m were carried out below 55 m depth, to initiate and stimulate new fractures and propagate them to

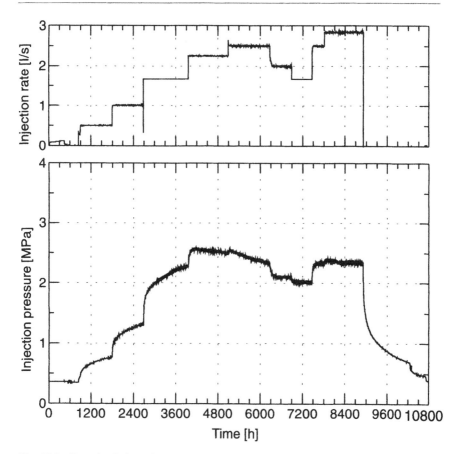

Fig. 13.4 Pre-stimulation injection test in borehole Waffenbrunn with an injection rate of up to 3 litres per second.

intersect water-bearing fractures in the gneiss rock mass. During each test a water volume of app. 1 m^3 was injected. To initiate fractures injection pressure values of more than 10 MPa were required. The conductivity of the rock within the test sections was increased from initially 10^{-10} m/s to almost 10^{-7} m/s (Fig. 13.5). Inspite of this, a single packer test at 80 m depth only showed an injectivity of 0.1×10^{-9} m^3/Pa s for the open-hole section below 80 m. As demonstrated by a final production test with a pump placed at 60 m depth, the production rate from the lower borehole section was at least 0.7 l/s (1.1×10^{-9} m^3/Pa s).

13.3.3 Borehole Lindau, N-Bavaia

The research borehole Lindau near Bayreuth was drilled to a depth of 530 m with a diameter of 146 mm. While the upper hundred metres were

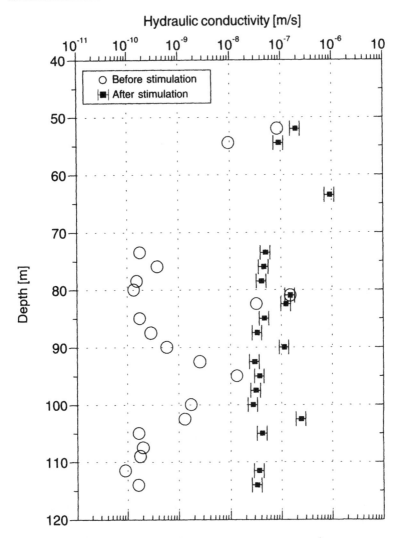

Fig. 13.5 Hydraulic conductivity of 3 m test sections in borehole Waffenbrunn before and after stimulation

cased, the borehole penetrates sandstones and conglomerates of the Lower Buntsandstone (to 178 m), then fine-grained sandstone of the Zechstone formation (to 281 m), and finally sandy claystones of the Upper Rotliegend formation. Early production tests showed a productivity of app. 4×10^{-9} m^3/Pa s with 97 per cent of the production originating from the Buntsandstone. After a massive hydraulic fracturing stimulation (300 m^3 of water) at app. 150 m, 165 m, and below 441 m depth the productivity could be increased to 6×10^{-9} m^3/Pa s.

To further investigate the possibility of productivity enhancement, 40 wireline hydrofracturing tests with a double straddle packer system of 3 m

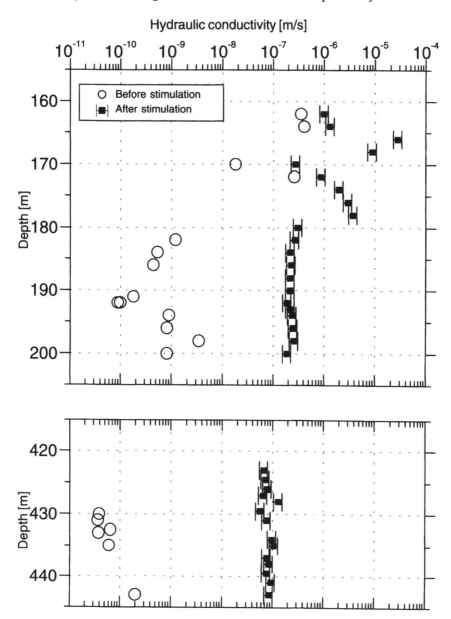

Fig. 13.6 Hydraulic conductivity of 3 m test sections in borehole Lindau before and after stimulation in Buntsandstone (162-178 m), Zechstone (178-200 m), and the Rotliegend formation (423-442 m).

test length were conducted, eight test in the rather permeable Buntsandstone, 14 tests in Zechstein sandstone, and 19 tests in the impermeable claystones of the Rotliegend formation below 418 m. The test series were carried out in such a way that a fracture was first induced at the bottom of the series and then extended zipper-like upwards with overlapping test intervals. At each test interval app. 100 litre water was injected into the induced fractures (or into the pore space of the permeable formation) at a rate of app. 0.5 l/s.

The results of this systematic investigation is shown in Fig. 13.6. The formation conductivity in the permeable Buntsandstone was increased by a factor of 16, the Zechstone conductivity by a factor of 220, and in the impermeable Rotliegend formation the conductivity increased by a factor of 860. However, the injectivity of the Zechstone/Rotliegendes still is 1 to 2 orders of magnitude smaller than the injectivity of the Buntsandstone. As shown by a final long-term production test the productivity of the borehole has only slightly increased from 4×10^{-9} or 6×10^{-9} to 8.6×10^{-9} m^3/Pa s.

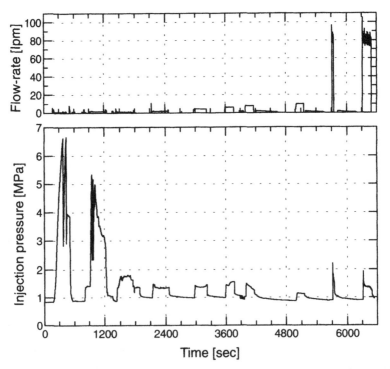

Fig. 13.7 Pressure and flow rate record during stimulation of the bottom hole section of borehole Herzog. Frac initiation at 7 MPa with an injection rate of 2 l/min, then fracture extension with high injection rates up to app. 100 l/min with negligible pressure response.

13.3.4   Borehole Herzog, Bochum, NW Germany

The local mineral water supplier has used an aquifer in the Upper Cretaceous Emscher claystone with boreholes drilled to app. 100 m since decades. To increase the supply one other borehole was drilled into the aquifer, but was essentially dry. The open-hole section from 72 m to 95 m had a diameter of 311 mm. An initial pulse test demonstrated a rock matrix permeability of 0.3 mDarcy. By application of the zipper-type hydraulic fracturing concept with overlapping test intervals, a fracture was induced in the bottom hole section at 92 m with a pressure of 7 MPa (Fig. 13.7) and then extended upwards to app. 80 m depth. The fracture extension pressure was only 2 MPa at injection rates up to 1.5 l/s. A total of 400 litres of water was injected during the stimulation operation. After pressure release no return flow was observed, an indication that the induced fracture had communication with a large aquifer. The final production test showed a constant production rate of 4.5 m^3 per day, the same yield as the neighbouring boreholes (Rummel, 1997).

13.4   CONCLUSIONS

For most areas in Central Europe a water well productivity of 10^{-9} m^3/Pa s (i.e. 2 l/s from a 300 m deep well) is not sufficient to significantly contribute to the drinking water consumption of a community. The situation could be different for rural areas in arid or semi-arid hard rock regions. The increase of the water yield from a well by a factor of 2 by hydraulic fracturing stimulation may not be economic but may be vital for the local population. As shown by the case histories, a productivity increase by factor of 2 is realistic.

By injection rates of some litres per minute, which are possible with a wireline straddle packer stimulation system, hydraulic fractures with an extension of deca-metres can be induced, however, the fracture width is much less than one millimetre. Shear offset on rough fracture planes is prevented by the shear resistance. Although large water quantities can flow through the fractures, they will almost close again after stimulation. Larger fracture widths can be obtained with high injection rates or using high viscosity fluids jacking the fracture open against the acting normal stress. Despite of the cost of such stimulation operations, the fracture jacking effect is mostly limited to the wellbore vicinity as shown by numerous laboratory studies (Klee, 19/1; Teza, 2002). The use of proppants in water well stimulation has not been systematically studied simply because of high costs. Treatment of induced fractures with acidic fluids is not appropriate because of environmental risks. Thus, the problem of effective water well stimulation in hard rock remains and is directly related to the local stress regime which controls fracture opening and closure. The situation will be different in water-bearing highly fractured rocks where hydrofracturing stimulation will create permanent intersection between the wellbore and the fracture water reservoir. A major fault zone may be considered as such an aquifer to be tapped by stimulation.

REFERENCES

Baria, R., Baumgärtner, J., Rummel, F., Pine, R.J., and Sato, Y., 1999. HDR/HWR reservoirs: concepts, understanding and creation. Geothermics, *28*:533-552. Pergamon/Elsevier Science.

Klee, G. 1991. Experimental investigation of the pressure distribution in hydraulic induced fractures. Yellow Rep. No. 4, Ruhr University Bochum, Inst. of Geophysics.

Rummel, F. 1997. Stimulation einer Mineralwasserbohrung mit geringer Schüttung. Der Mineralbrunnen 10:458-466.

Rummel, F. 2002. Crustal stress derived from fluid injection tests in boreholes. *In: In-situ characterization of rocks (eds. Sharma and Saxena), chapt. 6:* 205-244, Balkema Publishers.

Teza, T. 2001. The role of fractures as hydraulic valves in rocks. *Yellow Rep. No. 29, Ruhr University Bochum, Inst. of Geophysics.*

Thangarajan, M., Rai, S.N., and Singh, V.S., 2002. Sustainable development and management of groundwater resources in semi-arid regions with special reference to hard rocks. *Proc Int. Groundwater Conference.* Oxford & IBH Publ. Co., New Delhi.

Index

Abbreviations

a.s.l.	above sea level
ARB	accessible resource base
Bambus	backfill and material behaviour in underground salt repositories
BCM	billion cubic meters
BEM	boundary element method
CERI	Center for Earthquake Reserarch and Information
CMRI	Central Mining Research Institute
CMT	centroid moment tensor
CSIR	Centre of Scientific and Industrial Research
DAAD	German Academic Exchange Service
DOE	US Department of Energy
E	Young's modulus
EEIG	European Economic Interest Group
FDM	finite difference method
FEM	finite element method
g	gravity acceleration
G	Giga (10^9)
GAMMA	global positioning system array in Mid-America
G-P	Grassberger-Procaccio algorithm
GPS	global positioning system
H&W	Hubbert&Willis concept
HDR	hot dry rock concept/system/energy
HTPE	hydraulic test on pre-existing fractures
HZL	Hindustan Zinc Ltd., India
I	earthquake intensity
IUG	International Gas Union

J	Joule (energy unit)
k	kilo (10^3), permeability
KFA	today: Research Centre Jülich GmbH (FZ)
$k_{h/H}$	stress ratio between $S_{H,h}$ and S_v
K_I	stress intensity factor for mode I crack opening
K_{IC}	fracture toughness for mode I crack instability
LASL	Los Alamos Scientific Laboratory
LNG	liquified natural gas
LPG	liquified petroleum gas
M	Mega (10^6), earthquake magnitude
M_0	seismic moment
MMI	Modified Mercalli Intensity
Moho	Mohorovèiæ discontinuity
MRL	mean reference level
N	Newton (force unit)
NEIC	National Earthquake Information Center
NGRI	National Geophysical Research Institute Hyderabad, India
NMSZ	New Madrid Seismic Zone
NSF	US National Science Foundation
OECD	Organization for Economic Collaboration and Development
p, P	pressure
Pa	Pascal (pressure/stress unit)
p_C, P_C	critical pressure for fracture growth
p_{C0}, P_{C0}	hydraulic tensile strength
P_r	fracture re-opening pressure
PSI	shut-in pressure stress evaluation method
P_{si}	shut-in pressure, fracture closure pressure
RIS	reservoir induced seismicity
$S_{h,H}$	minor and major horizontal principal stress
S_n	normal stress component
SPIM	shut-in pressure inversion method
S_v	vertical stress
T	Tera (10^{12})
TBM	tunnel boring machine
TBS	testing borehole section

tCE	tons of coal equivalent
tOE	tons of oil equivalent
UGS	underground gas storage
URL	underground research laboratory
USGS	US Geological Survey
W	Watt (power unit)
WHFE	wireline hydraulic fracturing equipment
WIPP	waste isolation pilot plant
Θ	angle
μ	micro (10^{-6})
ν	Poisson's ratio
σ	normal stress
τ	shear stress
ω, Φ	phase angle
3DEC	three-dimensional distinct element code

Units

energy	1 J = 1 Joule	= 1 Joule = 1 Nm
	= 1 Ws	= 1 Watt/second
	1 kWh = 1 kilowatt/hour	
	1 GWh = 1 Gigawatt/hour	
	1 cal = 1 calorie = 4.18 J	
	1 tSKE = 11 tonne coal unit	

factors of magnitude

μ =	micro	= 10^{-6}
m =	milli	= 10^{-3}
k =	kilo	= 10^{3}
M =	mega	= 10^{6}
G =	Giga	= 10^{9}
T =	Tera	= 10^{12}
P =	Peta	= 10^{15}
E =	Exa	= 10^{18}

force   1 N = 1 Newton   = $1 \text{ kg} \cdot \text{m/s}^2$
               1 MN = 10^{6} N

fracture toughness $1 \text{ N/m}^{3/2} = 1 \text{ Pa} \cdot \text{m}^{1/2}$

heat flow 1 mW/m^2 = 1 milliwatt per square meter

hydraulic permeability $1 \text{ D} = 1 \text{ Darcy} \sim 10^{-12} \text{ m}^2$

moment   1 Nm = 1 Newton-meter

power   1 Nm = 1 Watt   = $1 \text{ m}^2 \cdot \text{kg/s}^3$

pressure/stress/   1 Pa = 1 Pascal   = 1 N/m^2
elastic moduli/   1 MPa = 10^{6} Pa

strength   1 GPa = 10^{9} Pa

viscosity   1 P = 1 Poise   = $1 \text{ Pa} \cdot \text{s}$

Printed and bound by CPI Group (UK) Ltd, Croydon, CR0 4YY

01/11/2024

01782636-0007